NEUROMETHODS ■ 39

Neuropeptide Techniques

NEUROMETHODS

Series Editor: *Wolfgang Walz*

NEUROMETHODS ■ 39

Neuropeptide Techniques

Edited by

Illana Gozes

Sackler School of Medicine, Tel Aviv University, Israel

HUMANA PRESS ✳ TOTOWA, NEW JERSEY

Production Editor: Rhukea J. Hussain

Cover design by Karen Schulz

Cover illustration: Green fluorescent receptor in cells in culture (Pilzer et al., this book); Receptor binding (Dangoor et al., this book); Western analysis of green fluorescent proteins (Holter et al., this book).

Printed in the United States of America. 10 9 8 7 6 5 4 3 2 1

eISBN 978-1-60327-099-1

Library of Congress Control Number: 2007936515

A Short Introduction

Illana Gozes, Ph.D.
Professor of Clinical Biochemistry
The Lily and Avraham Gildor Chair for the Investigation of Growth Factors
Director, The Adams Super Center for Brain Studies and The Edersheim Levie-Edesheim Gitter Institute for Functional Brain Imaging
Head, The Dr. Diana and Zelman Elton (Elbaum) Laboratory for Molecular Neuroendocrinology
Sackler Faculty of Medicine
Tel Aviv University
Tel Aviv 69978, Israel
Tel: 972-3-640-7240
Fax: 972-3-640-8541
E-mail: igozes@post.tau.ac.il

Chief Scientific Officer, Allon Therapeutics Inc. This book covers aspects of design, synthesis, and biological evaluation of peptides and peptide analogs. Emphasis is given to advanced technologies of transfection, transgenic delivery, and bioinformatics. Primary cells and cell lines are discussed as well as techniques in neuropeptide processing, trafficking, and secretion. Finally, *in vivo* techniques quantifying blood–brain barrier permeability to small proteins in mice and developmental milestones of behavior in newborn mice are described. The book gives a selected description of techniques in neuropeptide research written by leaders in the field.

Preface

Neuropeptide Methods: From Genes to Behavior and Back

Over the past 40 years, the field of neuropeptides has advanced from the discovery of peptides with biological action to the cloning and characterization of their respective precursor mRNA species and genes followed by identification of receptors. The major strides taken by the field of molecular biology, alternative RNA splicing, protein processing, and identification of gene control elements and transgenic animals have allowed a better understanding of neuropeptide production and function. Advanced organic chemistry has allowed the preparation of peptide analogs followed by peptidomimetics further deciphering the physiological impact of neuropeptides. G-protein coupled receptors that recognize many neuropeptides provide major targets for better disease diagnostics, and their ligands present important prototypes for lead discovery toward therapeutic drug development. Furthermore, structure-function activity studies also identify small peptide fragments in large proteins, providing leads for better therapeutics. Neuropeptides have an impact on multiple essential functions, from metabolic homeostasis to endocrine control and sexual function, from proper embryonic development and growth, to cancer diagnostics and therapy, pain management, cardiac and lung protection and neuroprotection.

The technological advances over the last decades are tremendous; this book provides the reader with broad perspectives and a breadth of knowledge on current topics related to neuropeptide research.

In particular, this book covers aspects of design, synthesis, and biological evaluation of peptides and peptide analogs. Emphasis is given to advanced technologies of transfection, transgenic delivery, and bioinformatics. Primary cells and cell lines are discussed as well as techniques in neuropeptide processing, trafficking, and secretion. Finally, *in vivo* techniques quantifying blood–brain barrier permeability to small proteins in mice and developmental milestones of behavior in newborn mice are described. The book gives a selected description of techniques in neuropeptide research written by leaders in the field.

I wish to thank all the contributors to this volume for their effort, and I am grateful to them for taking the time and having the patience to disseminate the detailed information required in order that others can succeed in the application of this technology in their practice. I am grateful to my excellent collaborators in science and in biotechnology over the years, who have inspired my scientific research. I would also like to thank Wolfgang Walz, the series editor, for support, and Patrick Marton and Gina Impallomeni for editorial assistance in organization and development of this volume.

Special thanks to Liat Nissanov, my dedicated administrative assistant, and to my immediate family, my husband, Dr. Yehoshua Gozes, our lovely daughter, Captain Adi Gozes, and my mother, Mrs. Esther Allon for their constant encouragement. In memory of my late father, Mr. Yizhak Allon, whose bright, honest, and kind spirit always guides me. To Mr. Thomas Lanigan, founder and president of Humana Press, who put me and many others on the track to excellent scientific editorship and publishing, I dedicate this book.

Illana Gozes
Tel Aviv, 2007

Contents

Contributors

Irina Arnaoutova • Section on Cellular Neurobiology, National
   Institute of Child Health and Human Development, National
   Institutes of Health, Bethesda, MD, USA
D.A. Carter • School of Biosciences, Cardiff University, Cardiff
Niamh X. Cawley • Section on Cellular Neurobiology, National
   Institute of Child Health and Human Development, National
   Institutes of Health, Bethesda, MD, UK
David Dangoor • Department of Human Molecular Genetics and
   Biochemistry, Sackler School of Medicine, Tel Aviv University,
   Department of Organic Chemistry, The Weizmann Institute
   of Science, Israel
J.S. Davies • School of Biosciences Cardiff University Cardiff, UK
Inna Divinski • Department of Human Molecular Genetics and
   Biochemistry Sackler School of Medicine, Tel Aviv University, Israel
Mati Fridkin • Department of Organic Chemistry, The Weizmann
   Institute of Science, Israel
Illana Gozes • Department of Human Molecular Genetics and
   Biochemistry, Sackler School of Medicine, Tel Aviv University, Israel
Joanna M. Hill • Laboratory of Behavioral Neuroscience, NIMH, NIH,
   Bethesda, MD, USA
J.L. Holter • Department of Basic Sciences and Aquatic Medicine, The
   Norwegian School of Veterinary Science, Ullevålsveien, Oslo,
   Norway
Miri Holtser-Cochav • Department of Human Molecular Genetic and
   Biochemistry, Sackler School of Medicine, Tel Aviv University, Israel
Sergy E. Ilyin • Johnson & Johnson Pharmaceutical Research
   & Development, L.L.C. Spring House, PA, USA
Robert T. Jensen • Digestive Diseases Branch, National Institute
   of Diabetes and Digestive and Kidney Diseases, NIH, Bethesda, MD,
   USA
Abba J. Kastin • Pennington Biomedical Research Center, Baton Rouge,
   LA, USA
Maria A. Lim • Laboratory of Behavioral Neuroscience, NIMH, NIH,
   Bethesda, MD, USA
Y. Peng Loh • Section on Cellular Neurobiology, National Institute
   of Child Health and Human Development, National Institutes
   of Health, Bethesda, MD, USA

Hong Lou • Section on Cellular Neurobiology, National Institute of Child Health and Human Development, National Institutes of Health, Bethesda, MD, USA

P.-S. Man • School of Biosciences, Cardiff University, Cardiff, UK

Terry W. Moody • Department of Health and Human Services, Center for Cancer Research, National Cancer Institute, NIH, Bethesda, MD, USA

Weihong Pan • Pennington Biomedical Research Center, Baton Rouge, LA, USA

Nimesh Patel • Section on Cellular Neurobiology, National Institute of Child Health and Human Development, National Institutes of Health, Bethesda MD, USA

Inbar Pilzer • Department of Human Molecular Genetics and Biochemistry, Sackler School of Medicine, Tel Aviv University, Israel

Carlos R. Plata-Salamán • Johnson & Johnson Pharmaceutical Research & Development, L.L.C. Spring House, PA, USA

Sara Rubinraut • Department of Organic Chemistry, The Weizmann Institute of Science, Israel

Michael Schumann • Digestive Diseases Branch, National Institute of Diabetes and Digestive and Kidney Diseases, NIH, Bethesda, MD, USA

Madeleine M. Stone • Laboratory of Behavioral Neuroscience, NIMH, NIH, Bethesda, MD, USA

Karin Vered • Department of Human Molecular Genetic and Biochemistry, Sackler School of Medicine, Tel Aviv University, Israel

T. Wells • School of Biosciences, Cardiff University, Cardiff, UK

Tulin Yanik • Section on Cellular Neurobiology, National Institute of Child Health and Human Development, National Institutes of Health, Bethesda, MD, USA

1

The Design, Synthesis, and Biological Evaluation of VIP and VIP Analogs

David Dangoor, Sara Rubinraut, Mati Fridkin, and Illana Gozes

Key Words: VIP; VPAC1; cAMP; HT29 cells; peptides; receptor binding.

Introduction

Peptides are currently considered as the new generation of biologically active tools despite the fact that they have been used for over a century to treat several kinds of diseases. Recent findings suggest a wide range of novel applications for peptides in medicine and biotechnology. The efficacy of native peptides has been greatly enhanced by introducing structural modifications in their original sequences.

Vasoactive intestinal peptide (VIP) is a linear 28 amino acid peptide that was first isolated from the duodenum and lung and characterized in the early 1970s by Mutt and Said *(1)*. It soon became clear that the functions and activities of this peptide are not limited to the gastrointestinal tract and to vasodilation. VIP is considered to be a prominent widely distributed neuropeptide in both the peripheral and the central nervous systems, where it functions as a nonadrenergic, noncholinergic neurotransmitter. The general physiological effects of VIP include vasodilatation, bronchodilation, anti-inflammatory actions, immunosuppression, cell proliferation, increase of gastric motility, and smooth muscle relaxation *(2)*. Clinical applications of VIP have been suggested before for male impotence, asthma, lung injury, diabetes, and a variety of tumors and neurodegenerative diseases *(2–4)*. The potential clinical applications of VIP have led to the thorough

From: *Neuromethods, Vol. 39: Neuropeptide Techniques*
Edited by: I. Gozes © Humana Press Inc., Totowa, NJ

characterization of this peptide and to the development of numerous VIP analogs *(5)*.

Multiplying the N-terminal of peptides is widely used in order to generate peptide derivatives with enhanced activities and stabilities such as the multiple antigen peptides (MAP) *(6)*. The design, synthesis, and purification methods of VIP analogs with multiple N-terminal domains will be described in this chapter. Furthermore, several biological assays that could be used to evaluate these peptides will be described.

Synthetic Procedure of VIP and VIP Analogs with Multiple N-Terminal Domains

All protected amino acids and coupling reagents were purchased from Novabiochem (Laufelfingen, Switzerland). Synthesis-grade solvents were obtained from Labscan (Dublin, Ireland). Peptides are prepared as C-terminal amides by conventional solid-phase peptide synthesis on rink amide resin, using an ABIMED AMS-422 automated solid-phase multiple peptide synthesizer (Langenfeld, Germany).

The 9-fluorenylmethoxycarbonyl (Fmoc) strategy is used throughout the peptide chain assembly *(7)*. Side-chain-protecting groups are *tert*-butyloxycarbonyl (*t*-Boc) for Lys; trityl (Trt) for Asn, Gln, and His; *tert*-butyl-ester (O-*t*-But) for Asp; *tert*-butyl ether (*t*-But) for Ser, Tyr, and Thr; and Pbf for Arg. Coupling is carried out in dimethylformamide (DMF) by using two successive reactions with 4 equivalents of the corresponding *N*-Fmoc amino acid, 4 equivalents of PyBOP reagent, and 8 equivalents of 4-methyl-morpholine, for 20–45 min at room temperature. Branched VIP analogs are prepared by the replacement of an amino acid along the VIP sequence with Lys, followed by the coupling of twofold equivalents of amino acid to the replaced Lys through its two amino groups (located α and $\acute{\varepsilon}$ to the carboxylic group of Lys). The twofold equivalents of amino acid that are coupled to the Lys are continuously coupled with two equivalents of amino acid until the first amino acid of the VIP sequence (His[1]). This methodology enables the attachment of two identical N-terminal sequences of the VIP to the replaced Lys in order to generate the branched VIP analogs (Table 1). Cleavage of the peptides from the polymer is performed by reacting the resin with trifluoroacetic acid (TFA)/H_2O/triethylsilane at volume ratios of 90/5/5 at room

Table 1. Amino Acid Sequence of Nle^{17}VIP and the Branched VIP Analog BR-2 and Their Mass Spectra Analysis

Peptide Name	Peptide Sequence	MH$^+$ Found (Calculated)
[Nle^{17}]VIP	H$_1$SDAV$_5$FTDNY$_{10}$ TRLRKQ[Nle]$_{17}$ AVKK-YLNSILN$_{28}$	3310.6 (3308)
Branched 2 (BR-2)	(HSDAVFT)$_2$Lys-[(Nle^{17}) VIP$_{9-28}$]	4080.3(4079)

temperature. The cleavage time from the resin is proportional to the number of Arg residues in the peptide, 2 h for each Arg. Crude peptides are purified by reversed-phase high-performance liquid chromatography (RP-HPLC) using a Vydac Protein/Peptide 218TP C-18 column (10 × 250 mm, 12-µm bead size; Vydac, Hesperia, CA, USA), employing a binary gradient formed from 0.1% TFA in water (solution A) and 0.1% TFA in 75% acetonitrile in water (solution B). The chromatographic run (flow rate 10 mL/min) starts with 10% of solution B in solution A, kept constant for 10 min, followed by a gradient increase of solution B from 10 to 100% over an additional 50 min. The isolated peptides are subjected to analytical RP-HPLC under similar conditions as those above to confirm their purity. Mass spectrometry is performed on a Micromass Platform LCZ 4000 (Manchester, UK) utilizing the electron spray ionization method. For amino acid composition analysis, peptides are hydrolyzed in 6 N HCl at 100 °C for 24 h under vacuum, and the hydrolyzates are analyzed with a Dionex automatic amino acid analyzer. All the VIP analogs are synthesized to include Nle^{17}, replacing the Met^{17}, as it was recently shown that this replacement does not change the peptide binding/activity toward the VPAC1 receptor *(8)* while enhancing its stability against oxidation.

The Expected Yield and Purity of the Synthesized Peptides

[Nle^{17}]VIP and the branched VIP analogs are obtained in high yields (80–85%). The peptide purity (assessed by RP-HPLC) is higher than 97%. Representative mass spectra data of the HPLC-purified [Nle^{17}]VIP and the branched VIP analog BR-2 are given in Table 1 and are in agreement with the expected mass. Amino acid ratios are found to be rather identical to the expected values (data not shown).

Binding Assay on HT-29 Cells

Human colonic HT-29 cells expressing only the VPAC1 receptor *(9)* are chosen for the analysis of the peptides receptor binding according to previously reported conditions *(8,10)*. HT-29 cells are routinely cultured in 75-cm^2 culture flasks in Dulbeco's modified Eagle's mdium (DMEM) containing 4.5 g/L glucose supplemented with 10% (V/V) fetal calf serum (FCS) and penicillin/streptomycin (100 U/mL and 0.1 mg/mL, respectively) in a humidified atmosphere of air/CO_2 (95%/5%) at 37 °C. The culture medium is replaced by fresh medium every 3 days. For subcultures, cells are harvested in versene for 5 min at 37 °C. Cells (4 × 10^5 cells/well) are seeded in collagen precoated 24 well plates and cultured for 2 days. The cells are pre-incubated for 1 h at 4 °C and then incubated for 3 h at 4 °C in the presence of 50 pM ^{125}I-VIP (Amersham 2200 Ci/mmol) and increasing concentrations of VIP or VIP analogs in DMEM-50 mM HEPES (pH = 7.4) containing 1% bovine serum albumin (BSA), 0.1% bacitracin, and 150 µM phenylmethylsufonylfluoride (PMSF). Binding reactions are stopped by cooling dishes on ice. Cells are rinsed once with 2 mL of cold phosphate buffer saline (PBS) and lysed in 400 µL of 0.5 N NaOH. Radioactivity in cell lysates is quantified in a gamma counting system. Specific binding is calculated as the difference between the amount of ^{125}I-VIP bound in the absence (total binding) and the presence of 1 µM unlabeled VIP (nonspecific binding).

The Peptides' Expected IC$_{50}$ Values Following Binding Assay on HT-29 Cells

Data obtained from binding experiments allow the determination of the IC$_{50}$, the concentration of unlabeled peptide leading to half-maximal inhibition of the ^{125}I-VIP specific binding. [Nle^{17}]VIP potently (Fig. 1) inhibits specific ^{125}I-VIP binding to HT-29 cells. The IC$_{50}$ obtained for [Nle^{17}]VIP (IC$_{50}$ = 7 × 10^{-10} M) on HT-29 cells is sixfold lower than that of the branched VIP analogs, BR-2 (IC$_{50}$ = 4.5 × 10^{-9}).

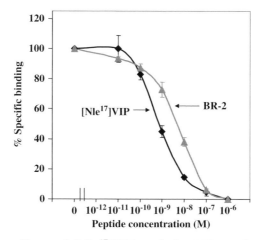

Fig. 1. Dose effects of [Nle^{17}]VIP and the VIP analog BR-2 for the inhibition of ^{125}I -VIP binding to HT29 cells. Cells (4 × 10^5 cells/well) are seeded on collagen precoated 24 well plates and cultured for 2 days. Incubations with radioligand (50 pM, ^{125}I VIP) is performed at 4 °C for 180 min in the presence of an increasing concentration of unlabeled peptides. Nonspecific binding is subtracted for each value. Data are the means (±SEM) of at least two independent triplicate experiments.

Intracellular cAMP Formation Assay

HT-29 cells (2 × 10^5) are seeded in 24-well dishes and cultured for 3 days, after which their number is determined in three wells. The culture medium is then removed, and cells are washed once with 500 µL of fresh medium and then equilibrated at 37 °C with 500 µL of medium containing 0.1 mM 3-isobutyl methyl xantine (IBMX) for 30 min. Cells are then incubated for 30 min at 37 °C after addition of the peptides. Control cells are treated with saline. The medium is then removed, and the cAMP intracellular content is determined via the use of ELISA kit (Amersham).

The Effects of VIP and VIP Analogs on Intracellular cAMP Levels

The effects of [Nle^{17}]VIP and VIP analogs on cAMP intracellular levels are assayed following an incubation time of 30 min with the peptides, in dose-response experiments (Fig. 2). [Nle^{17}]VIP is highly efficient in elevating cAMP basal levels (~80-fold increase as compared to the control). The EC$_{50}$ value (concentration required

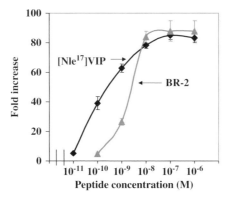

Fig. 2. Dose-dependent effects of [Nle^{17}]VIP and the branched VIP analog BR-2 on intracellular cAMP levels in HT-29 cells. Data are the means of at least two independent triplicate experiments. The control cells are treated with saline. The results are presented as the fold increase relative to the control cells. The control cells level of cAMP is 1.5 pmol/10^6 cells.

to achieve 50% of the maximal effect) obtained by [Nle^{17}]VIP (EC$_{50}$=1 × 10^{-10}M) is in accord with its binding, IC$_{50}$ = 7 × 10^{-10}M. The VIP analog BR-2 demonstrates reduced potency as compared to [Nle^{17}]VIP (20-fold lower than [Nle^{17}]VIP, respectively; Fig. 2). This is in accordance with its lower affinity toward the VIP receptor as compared to the [Nle^{17}]VIP (six-fold lower than [Nle^{17}]VIP).

Evaluation of the Peptides' Secondary Structure: Circular Dichroism Studies

Circular dichroism (CD) analysis is performed in order to elucidate the peptides' secondary structure. CD spectra are recorded on an AVIV-202 circular dichroism spectrometer (Lakewood, NJ, USA). Duplicate scans over a wavelength range of 190–260 nm are taken at ambient temperature. Peptides are dissolved in TFE (trifluroethanol)/DDW (40/60, V/V) at a final concentration of 0.05 mM. A baseline is recorded and subtracted after each spectrum. Ellipticity is reported as the mean residue ellipticity [ϕ] in degrees-cm^2-dmol^{-1}. [ϕ]= [ϕ]$_{OBS}$(MRW/10 l C), [ϕ]$_{OBS}$ is the ellipticity measured in millidegrees, MRW is the mean residue molecular weight of the peptide (molecular weight divided

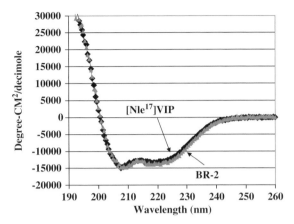

Fig. 3. Circular dichroism spectra of [Nle^{17}]VIP and the branched VIP analog BR-2 in 40% TFE. Ellipticity is reported as the mean residue ellipticity [φ] in degrees-CM^2-dmol^{-1}. [φ] = [φ]$_{OBS}$(MRW/10 l C). [φ]$_{OBS}$ is the ellipticity measured in millidegrees, MRW is the mean residue molecular weight of the peptide (molecular weight divided by the number of residues), C is the concentration of the sample in mg/mL, and l is the optical path length of the cell in cm.

by the number of residues), C is the concentration of the sample in mg/mL, and l is the optical path length of the cell in cm.

The Peptides' Predicted Secondary Structure

The CD spectra of [Nle^{17}]VIP in 40% TFE (Fig. 3) reveal two negative minima, at 222 nm and at 208 nm, indicating an α helical structure. This is in agreement with previous studies *(11,12)* that demonstrated that the central part of VIP adopts an alpha helical structure in an organic environment. The CD spectrum of the VIP analog BR-2 in 40% TFE (Fig. 3) appears highly similar to that of the [Nle^{17}]VIP, with two minima, at 222 nm and at 208 nm, indicating an alpha helical structure.

Evaluation of the Metabolic Stability of VIP Analogs: Trypsin-Catalyzed Cleavage of VIP and VIP Analogs

[Nle^{17}]VIP or VIP analogs are dissolved in 115 μL of 0.05 M HEPES buffer, pH = 7.4 (100 μM final peptide concentration), and incubated with 10 μL of trypsin (1 μg/μL) at 37 °C. Ten μL of

1 M HCL are added at different time points in order to stop the reaction. HPLC analysis is performed on all the samples, and the percent of undigested peptide at each time point is determined. The peptides' degradation products following incubation with trypsin are characterized by amino acid analysis.

Expected Results Following [Nle^17]VIP Incubation with Trypsin

HPLC analysis of the crude mixture of [Nle^{17}]VIP and trypsin following 30 minutes' incubation reveals five major peaks: One peak stands for the [Nle^{17}]VIP and four peaks correspond to [Nle^{17}]VIP's degradation products. These peaks are examined by amino acid analysis, and the following degradation products are identified: **(1)**VIP$^{1-12}$, **(2)**VIP$^{1-14}$, **(3)**VIP$^{13-28}$, **(4)**VIP$^{15-28}$. Hence, the results are consistent with cleavage of VIP at two primary sites: the C-terminal side of Arg^{12}, and the C-terminal side of Arg^{14}. This is accordance with trypsin's known activity to cleave peptides at the C-terminal side of Lys and Arg. During the incubation time of [Nle^{17}]VIP in trypsin, [Nle^{17}]VIP is gradually digested as determined by the HPLC analysis (Fig. 4). $T_{1/2}$, the time required to degrade 50% of the [Nle^{17}]VIP, is 30 min (Fig. 4). No significant degradation of VIP was observed in incubation over the same interval in the absence of trypsin.

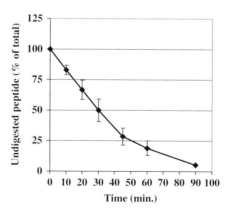

Fig. 4. Enzymatic digestion of [Nle^{17}]VIP by trypsin. At the indicated period, the relative quantity of the intact [Nle^{17}]VIP in the reaction mixture was determined by HPLC analysis. Each point represents the percentage (mean ± SD, n = 4) of undigested peptide.

References

1. Said, S.I., and Mutt, V. A peptide fraction from lung tissue with prolonged peripheral vasodilator activity. *Scand. J. Clin. Lab. Invest.*, 1969; **107**: 51–56.
2. Gozes, I., and Furman, S. VIP and drug design. *Curr. Pharm. Des.*, 2003; **9**(6): 483–494.
3. Said, S.I. Vasoactive intestinal polypeptide biologic role in health and disease. *Trends Endocrin. Metab.*, 1991; **2**(3): 107–112.
4. Gozes, I., et al. Pharmaceutical VIP: Prospects and problems. *Curr. Med. Chem.*, 1999; **6**(11): 1019–1034.
5. Gozes, I., et al. Stearyl-norleucine-vasoactive intestinal peptide (VIP): A novel VIP analog for noninvasive impotence treatment. *Endocrinology*, 2004; **134**(5): 2121–2125.
6. Bracci, L., et al. A branched peptide mimotope of the nicotinic receptor binding site is a potent synthetic antidote against the snake neurotoxin alpha-bungarotoxin. *Biochemistry*, 2002; **41**: 10194–10199.
7. Atherton, E., and Sheppard, R.C. *Solid Phase Peptide Synthesis. A Practical Approach*. IRL Oxford University Press, New York, 1989.
8. Dangoor, D., et al. Novel extended and branched N-terminal analogs of VIP. *Regul. Pep.*, 2006; **15**;137(1–2): 42–49.
9. Summers, M.A., et al. A lymphocyte-generated fragment of vasoactive intestinal peptide with VPAC1 agonist activity and VPAC2 antagonist effects. *J. Pharmacol. Exp. Ther.*, 2006; **306**(2): 638–645.
10. Lelievre, V., et al. Differential expression and function of PACAP and VIP receptors in four human colonic adenocarcinoma cell lines. *Cell Signal*, 1998; **10**(1): 13–26.
11. Fournier, A., et al. Conformational analysis of vasoactive intestinal peptide and related fragments. *Ann. NY Acad. Sci.*, 1988; **527**: 51–67.
12. Fry, D.C., et al. Solution structure of an analogue of vasoactive intestinal peptide as determined by two-dimensional NMR and circular dichroism spectroscopies and constrained molecular dynamics. *Biochemistry*, 1989; **28**(6): 2399–2409.

2

Antibody Production: Activity-Dependent Neuroprotective Protein (ADNP) as an Example

Illana Gozes, Miri Holtser-Cochav, and Karin Vered

Abstract

Activity-dependent neuroprotective protein (ADNP, human calculated molecular mass 123,562.8 Da) is a newly discovered glial protein that it is essential for embryonic development and brain formation. ADNP includes an active neuroprotective site, an 8 amino acid peptide NAPSVIPQ (NAP). The current study was set out to prepare antibodies to ADNP that will recognize different sites on the molecule. Four peptides of 8–20 amino acids that span the ADNP molecule, including NAPVSIPQ, were prepared. Peptides (containing a Cys residue attached to the N-terminal amino acid) were conjugated to keyhole limpet hemocyanin (KLH) and injected to respective rabbits in the presence of Freund's complete adjuvant. Following five booster injections in incomplete Freund's adjuvant, the respective antisera were collected and assayed by enzyme-linked immunosorbent assay (ELISA) and purified by affinity chromatography on peptides conjugated to SulfoLink Coupling Gel (Pierce). Mouse brain proteins were prepared (4 months old) and separated into cytoplasmic and nuclear fractions. Proteins were further separated by SDS-PAGE (SDS-PolyAcrylamide Gel Electrophoresis) and transferred to nitrocellulose membranes. Membranes were then probed with the antibodies followed by a secondary antibody horseradish peroxidase conjugated anti-rabbit IgG prepared in goat and developed with ECL. All antibodies recognized intact ADNP in both cytoplasmic and nuclear fractions. This work developed antibodies against ADNP and NAP that will be utilized for further experimentations to elucidate the distribution and mechanisms of ADNP and NAP neuroprotection.

Introduction

Activity-dependent neuroprotective protein (ADNP) is a gene product essential for brain formation (1,2) that is regulated, in part, by the neuropeptide vasoactive intestinal peptide (VIP). The

From: *Neuromethods, Vol. 39: Neuropeptide Techniques*
Edited by: I. Gozes © Humana Press Inc., Totowa, NJ

ADNP gene is highly conserved in the mouse *(3)*, in humans *(4)*, and in the rat *(5)* and is abundantly expressed in the brain and the body *(3,4)*. Recombinant ADNP provides neuroprotection *(6)*. Peptide activity scanning identified NAP (NAPVSIPQ) as a small active fragment of ADNP that provides neuroprotection at very low concentrations. In cell culture, NAP has demonstrated protection against toxicity associated with the beta-amyloid peptide, N-methyl-D-aspartate, electrical blockade, the envelope protein of the AIDS virus, dopamine, H_2O_2, nutrient starvation, and zinc overload. NAP has also provided neuroprotection in animal models of apolipoprotein E deficiency, cholinergic toxicity, closed head injury, stroke, middle-aged anxiety, and cognitive dysfunction. NAP interacts with tubulin and facilitates microtubule assembly, leading to enhanced cellular survival that is associated with fundamental cytoskeletal elements (for a review, see *(7)*). To study the function of any given protein, it is essential to prepare a set of antibodies that allow localization, quantitation, and functional assessments. Previously, antibodies were prepared for ADNP characterizing expression and localization *(4–9)*. Here, the methodology is explained for further studies.

Materials and Methods

Antigen Preparation and Immunizations

The choice of peptides for antibody production is outlined in Fig. 1. Four peptides (containing a Cys residue attached to the N-terminal amino acid, to allow S-S bond formation) were synthesized as described in this book. In order to enhance the immune response against a short peptide (representing a few antigenic determinants), the peptide antigen is conjugated to a carrier molecule. An example for a carrier protein is keyhole limpet hemocyanin (KLH; MW 4.5×10^5 to 1.3×10^7 Dalton). One mg of KLH is diluted in 1 mL of PBS, and 200 µL of this solution are added to 4 mg peptide/mL of PBS. The peptide-KLH mixture is then incubated for 2 h at ambient temperature. To confirm conjugation, a DTNB, Ellman's reagent (5,5'-Dithio-bis(2-nitrobenzoic acid) test, is performed: Five µL of the conjugated peptide are added to 50 µL of PBS and 5µL of 1 mg/mL DTNB. The mixture is incubated for 2–3 h: If the solution does not turn yellow, the peptide-KLH conjugation mixture is further incubated for another 12 h at 4 °C and a second DTNB test is performed.

```
MFQLPVNNLGSLRKARKTVKKILSDIGLEYCKEHIEDFKQFEPNDFYLKNTTWEDVGLWD
PSLTKNQDYRTKPFCCSACPFSSKFFSAYKSHFRNVHSEDFENRILLNCPYCTFNAD
KKTLETHIKIFHAPNSSAPSSSLSTFKDKNKNDGLKPKQADNVEQAVYYCKKCTYRD
PLYEIVRKHIYREHFQHVAAPYIAKAGEKSLNGAVSLGTNAREECNIHCKRCLFMPK
SYEALVQHVIEDHERIGYQVTAMIGHTNVVVPRAKPLMLIAPKPQDKKGMGLPPRIS
SLASGNVRSLPSQQMVNRLSIPKPNLNSTGVNMMSNVHLQQNNYGVKSVGQSYGVGQ
SVRLGLGGNAPVSIPQQSQSVKQLLPSGNGRSFGLGAEQRPPAAARYSLQTANTSLP
PGQVKSPSVSQSQASRVLGQSSSKPPPAATGPPPSNHCATQKWKICTICNELFPENV
YSVHFEKEHKAEKVPAVANYIMKIHNFTSKCLYCNRYLPTDTLLNHMLIHGLSCPYC
RSTFNDVEKMAAHMRMVHIDEEMGPKTDSTLSFDLTLQQGSHTNIHLLVTTYNLRDA
PAESVAYHAQNNAPVPPKPQPKVQEKADVPVKSSPQAAVPYKKDVGKTLCPLCFSIL
KGPISDALAHHLRERHQVIQTVHPVEKKLTYKCIHCLGVYTSNMTASTITLHLVHCR
GVGKTQNGQDKTNAPSRLNQSPGLAPVKRTYEQMEFPLLKKRKLEEDADSPSCFEEK
PEEPVVLALDPKGHEDDSYEARKSFLTKYFNKQPYPTRREIEKLAASLWLWKSDIAS
HFSNKRKKCVRDCEKYKPGVLLGFNMKELNKVKHEMDFDAEWLFENHDEKDSRVNAS
KTVDKKHNLGKEDDSFSDSFEHLEEESNGSGSPFDPVFEVEPKIPSDNLEEPVPKVI
PEGALESEKLDQKEEEEEEEEDGSKYETIHLTEEPAKLMHDASDSEVDQDDVVEWK
DGASPSESGPGSQQISDFEDNTCEMKPGTWSDESSQSEDARSSKPAAKKKATVQDDT
EQLKWKNSSYGKVEGFWSKDQSQWENASENAERLPNPQIEWQNSTIDSEDGEQFDSM
TDGVADPMHGSLTGVKLSSQQA
```

Fig. 1. ADNP sequence and derived peptides. Mouse ADNP-amino acid sequence *(3)* and the sites of the four peptides on the molecule are either highlighted in grey or underlined.

The peptide-KLH conjugate is then purified by dialysis. Dialysis cassettes (Slide-A-Lyzer, Pierce) are dipped in PBSx1 for 30 s, and the conjugated peptide is then injected to the dialysis cassette, followed by a 12-h incubation period in PBS, at 4 °C. The conjugated peptide is then collected from the dialysis cassette, diluted with PBS to a volume of 2 mL, and mixed with 2 mL of complete Freund's adjuvant.

The antigen/Freund's adjuvant emulsion is then injected SC into four different New Zealand rabbits after collection of the respective pre-immune serum. For the five following booster injections (with a 3-week interval), the conjugated peptides are emulsified with incomplete Freund's adjuvant. Serum collection is performed monthly.

The procedure described above is general and was used was antibody preparation against ADNP.

Evaluating an Immune Response

Dot-Blot Analysis

The serum collected after the second booster was analyzed by dot-blot in order to detect the presence of antibodies. As control we used pre-immune serum collected from the four rabbits.

The respective peptides were applied onto nitrocellulose membranes and dried for 45 min. The membranes were incubated with a blocking solution (10 mM Tris, 6 mM NaCl, 0.05% Tween-20, and 10% lowfat milk) for 1 h, followed by incubation with the respective serum/pre-immune serum (1:500) for 16 h. Membranes were then incubated with horseradish peroxidase (HRP) conjugated goat anti-rabbit IgG (1:500)for 1 h. After washing, the membranes were developed with 100 µL HRP substrate, tetramethylbenzidine (DAKO), and reactions were stopped with 100 µL 1 M H_2SO_4. All analyses were performed in triplicate. Reactions were developed with ECL+ (Western blotting detection system, Amersham Pharmacia Biotech, Buckinghamshire, UK).

Enzyme-Linked Immunosorbent Assays (Specificity of Antibodies Against the Four Peptides)

Enzyme-linked immunosorbent assays (ELISA) were performed in order to check for the specificity of the antibodies. Ninety-six-well immunoassay plates (Maxisorp; Nunc) were coated overnight at 4 °C with 50 µL/well of 1 µg of peptide/mL of 0.05 M bicarbonate buffer at pH 9.6. Wells were then washed with 50 µL/well of PBS and then blocked for 1 h at 37 °C with 50 µL/well of PBS + 2% BSA,and 0.05% Tween 20 (PBT). Serum samples were diluted (1:100,000,000) in PBS-M–0.1% (vol/vol) Tween 20 (PBS-MT) and added to the plates (50 µL/well). Plates were then incubated at 37 °C for 1 h, washed 10 times with PBS–0.1% Tween 20 (PBS-T), and probed for 1 h at 37 °C with a secondary antibody conjugated to alkaline phosphatase (1:1000) and incubated for 1 h at 37 °C or room temperature. Then a pNPP alkaline phosphatase substrate (Sigma) was added and incubated for 30 min at room temperature. Negative control reactions were performed by coating with native ovalbumin and probing as described above. The plate was read at a wavelength of 405 nm.

Purification of Antibodies from Serum by Affinity Chromatography

Serum after the fifth boost was collected and used. In order to purify antibodies from the serum, two different affinity chromatography columns were used: *(1)* ImmunoPure (A Plus) IgG purification; *(2)* sulfolink-coupling gel.

At first, serum was loaded on the ImmunoPure (A Plus) IgG purification column. Protein A-binding IgG fractions were eluted with an elution buffer according to manufacturer's instructions (Pierce).

The second isolation efforts utilized a different affinity column, sulfolink-coupling gel (Pierce). Binding of the four peptides to the respective columns was performed according to the manufacturer's instructions. Sera were then loaded on the respective columns, and the peptide-binding IgG fractions were eluted with 0.1 M glycine-HCl.

Mouse brain proteins were prepared (4 months old) and separated into cytoplasmic and nuclear fractions. Proteins were further separated by SDS-PAGE (SDS-polyacrylamide gel electrophoresis) and transferred to nitrocellulose membranes. Membranes were then probed with the antibodies followed by a secondary antibody horseradish peroxidase conjugated goat antirabbit IgG and developed with ECL.

Results

Identification of Antibodies' Presence in the Serum After the Second Booster Using Pre-immune Serum as a Control

Serum produced from the blood collected prior to the primary immunization and after the second booster was analyzed by dotblot. Each dot on the blot contained 1 µg of the peptide. Each blot was examined with one of the pre-immune serum samples at a 1:500 dilution, and the serum collected after the second booster was examined with the appropriate peptide blot. A secondary antibody (goat anti-rabbit IgG, 1:25,000 dilution) was added for 1 h at 25 °C. The reaction was developed using the ECL kit (Amersham Biosciences).

The pre-immune serum did not show immune detection of the four peptides in contrast to the serum samples examined after the second booster (Fig. 2).

Antibodies Against the Four Peptides Recognize These Peptides

Serum produced from blood collected after the fifth booster was analyzed by enzyme-linked immunosorbent assay, using the pre-immune serum as a control. Ninety-six- well plates were

(a)

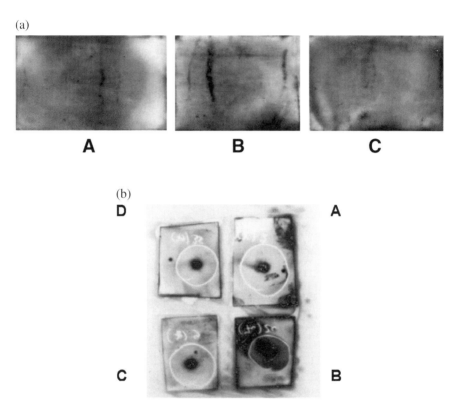

Fig. 2. (a) The pre-immune serum did not show immunodetection of the peptides. A-CYREHFQHVAAPYIAK ,B-CLGLGGNAPVSIPQQ, C-CNAPVSIPQQSQSVKQLLPS. (b)The serums after the second boost showed immunodetection of the four peptides. A-CNAPVSIP QQSQSVKQLLP, B-CYREHFQHVAAPYIAK, C-CPMHGSLAGVKL SSQQA, D-CLGLGGNAPVSIPQQ.

coated with one of the four tested peptides. The serum/pre-immune serum samples were diluted 1:100–1:10^6 as described in the methods section.

Results showed activity at serum dilutions of 1:100, 1:1000, and sometimes at 1:10,000 (antibodies against CLGLGGNAPVSIPQQ). The background activity of the pre-immune serum samples was high (see Fig. 3).

An Example for Anti-CYREHFQHVAAPYIAK Purification

Cytoplasmic and nuclear proteins were extracted from mouse brains *(10)*. The extracts were loaded on 10% polyacrylamide

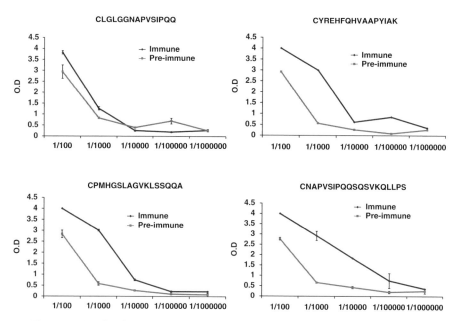

Fig. 3. Antibody reaction against the four peptides. Comparison between pre-immune serums and the serums taken after the fifth booster, for each peptide antigen.

gels and separated by size. Western blot analysis was performed (4) using (1) serum after the fifth booster, (2) antibodies eluted from the ImmunoPure (A Plus) IgG purification column, and (3) antibodies eluted from the sulfo-link coupling gel column. Figure 4 shows that affinity chromatography seems to yield purer antibodies against the peptide, with increased recognition of bands at the ~ 100,000 Da region, the expected ADNP size. Figure 5 compares the affinity-purified antibodies to the four peptides. There seems to be an apparent difference between antibodies recognizing the C-terminal vs. the N-terminal of ADNP.

Specificity and Conclusions

To evaluate antibody specificity, antibodies prepared against a certain peptide were reacted against another in an ELISA assay. An example is shown for the antibody against CPMHGSLAGVKLSSQQA that does not react with CYREHFQH-VAAPYIAK, and vice versa (see Fig. 6). This short chapter describes the use of peptide antigens to prepare antibodies again a protein.

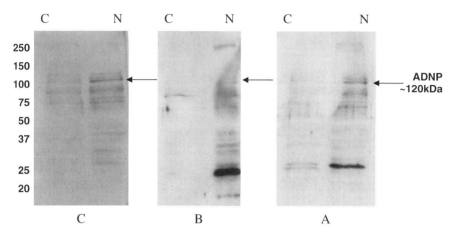

Fig. 4. An example for purification of the antibody against the peptide CYREHFQHVAAPYIAK. An example of antibody enrichment (Western blot) of brain cytoplasmic [C] and nuclear [N] fractions with antibody against the peptide CYREHFQHVAAPYIAK). (A) Total serum. (B) Purification on ImmunoPure (A Plus) IgG purification column. (C) Purification on sulfo-link gel—affinity chromatography.

These antibodies show specificity and may be developed to use for antigen quantitation. ADNP is an essential protein that is associated with the expression of hundreds of key genes *(1)*. NAPVSIPQ that is derived from ADNP (Fig. 1) is a key neuroprotective peptide in clinical development *(7)*. Antibodies are important research tools for the evaluation of ADNP and NAP, e.g., *(4,8–10)*.

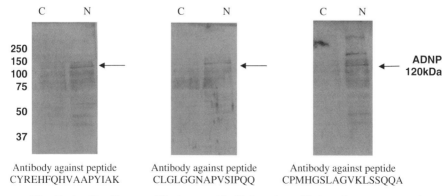

Fig. 5. Antibodies purified on sulfo-link gel—affinity chromatography. All antibodies recognized intact ADNP at 120 kD in both cytoplasmic (C) and nuclear (N) fractions.

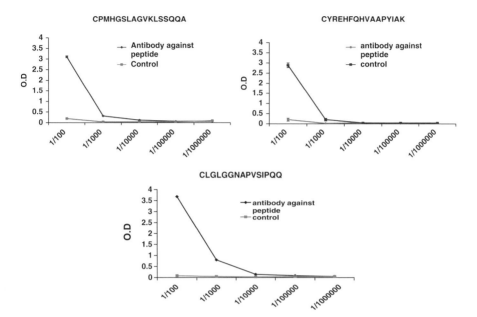

Fig. 6. Specificity of antibodies. ELISA analysis evaluated the specificity of the antibodies by cross checking. Antibodies prepared against one peptide do not react with the other peptides. An example is shown for antibody against CPMHGSLAGVKLSSQQA that does not react with CYREHFQHVAAPYIAK, and vice versa (an antibody dilution curve is shown).

Acknowledgments

This study was supported, in part, by the U.S.-Israel Binational Science Foundation, the Israel Science Foundation, and Allon Therapeutics Inc. Professor Illana Gozes is the incumbent of the Lily and Avraham Gildor Chair for the Investigation of Growth Factors at Tel Aviv University and the director of the Adams Super Center for Brain Studies and the Levie-Edersheim-Gitter fMRI Institute, and the Dr. Diana and Zelman Elton (Elbaum) Laboratory for Molecular Neuroendocrinology. Professor Illana Gozes serves as the chief scientific officer of Allon Therapeutics Inc.

References

1. Mandel, S., Rechavi, G., and Gozes, I. Activity-dependent neuroprotective protein (ADNP) differentially interacts with chromatin to regulate genes essential for embryogenesis. *Dev. Biol.*, 2007; **303**(2): 814–824.

2. Pinhasov, A., et al. Activity-dependent neuroprotective protein: A novel gene essential for brain formation. *Brain Res. Dev. Brain Res.*, 2003; **144**(1): 83–90.

3. Bassan, M., et al. Complete sequence of a novel protein containing a femtomolar-activity-dependent neuroprotective peptide. *J. Neurochem.*, 1999; **72**(3): 1283–1293.

4. Zamostiano, R., et al. Cloning and characterization of the human activity-dependent neuroprotective protein. *J. Biol. Chem.*, 2001; **276**(1): 708–714.

5. Sigalov, E., et al. VIP-related protection against Iodoacetate toxicity in pheochromocytoma (PC12) cells: A model for ischemic/hypoxic injury. *J. Mol. Neurosci.*, 2001; **15**(3): 147–154.

6. Steingart, R.A., and Gozes, I. Recombinant activity-dependent neuroprotective protein protects cells against oxidative stress. *Mol. Cell. Endocrinol.*, 2006; **252**(1–2): 148–153.

7. Gozes, I., et al. NAP: Research and development of a peptide derived from activity-dependent neuroprotective protein (ADNP). *CNS Drug Rev.*, 2005; **11**(4): 353–368.

8. Gozes, I., et al. The expression of activity-dependent neuroprotective protein (ADNP) is regulated by brain damage and treatment of mice with the ADNP derived peptide, NAP, reduces the severity of traumatic head injury. *Curr. Alzheimer Res.*, 2005; **2**(2): 149–153.

9. Furman, S., et al. Sexual dimorphism of activity-dependent neuroprotective protein in the mouse arcuate nucleus. *Neurosci. Lett.*, 2005; **373**(1): 73–78.

10. Furman, S., et al. Subcellular localization and secretion of activity-dependent neuroprotective protein in astrocytes. *Neuron Glia Biol.*, 2004; **1**(3): 193–199.

3

Primary Cell Cultures and Cell Lines

Inna Divinski, Inbar Pilzer, and Illana Gozes

Key Words: tissue culture; cell lines; subculture; primary cells.

Introduction

Animal cells have been cultivated *in vitro* since the 1900s. Cell culture offers a means to study cellular responses *in vitro* under controlled conditions and is increasingly used due to international ratification for the "3Rs" principle—i.e., reduction, refinement, and replacement of the use of animals for *in vitro* experimentation. Cells for *in vitro* culture may be derived directly from tissue as primary cells or may be available as cell lines that can be subcultured and prepared as cell banks *(1)*.

Animal and human cells may be studied *in vitro* by a variety of approaches using whole organs or pieces of tissue as well as isolated cells and cell lines. However, this chapter will focus on the use of glial and neuronal cultures derived from newborn rats' brain and cell lines.

Rat Cerebral Cortical Astrocyte Cell Cultures

Newborn rats (Harlan, Jerusalem, Israel) are sacrificed by decapitation, and the brain is removed *(2)*. The cortex is dissected, and the meninges are removed. The tissue is minced with scissors and placed in Hank's balanced salts solution 1 (S1), containing HBSS (Biological Industries, Beit Haemek, Israel), 15 mM HEPES buffer, pH 7.3 (Biological Industries, Beit Haemek, Israel), and 0.25% trypsin (Biological Industries) in an incubator at 37 °C 10% CO_2 for 20 min. The cells are then placed in 5 mL of solution 2 (S2) containing 10% heat inactivated fetal calf

From: *Neuromethods, Vol. 39: Neuropeptide Techniques*
Edited by: I. Gozes © Humana Press Inc., Totowa, NJ

serum (Biological Industries), 0.1% gentamycin sulphate solution (Biological Industries), and 0.1% penicillin-streptomycin-nystatin solution (Biological Industries) in Dulbecco's modified Eagle's medium (DMEM, Sigma, Rehovot, Israel) . The cells are allowed to settle and are then transferred to a new tube containing 2.5 mL of S2 and triturated using a Pasteur pipette. The process is repeated twice more. Once all the cells are suspended, cell density is determined using a hemocytometer (Neubauer Improved, Germany) and 15×10^6 cells/15 mL S2 are inoculated into each 75-cm^2 flask (Corning, Corning, NY, USA). Cells are incubated at 37 °C 10% CO_2. The medium is changed after 24 h, and cells are grown until confluent.

Rat Cerebral Cortical Astrocytes Cell Subcultures

The flasks are shaken to dislodge residual neurons and oligo-dendrocytes that may be present. Flasks are then washed with 10 mL cold HBSSx1, HEPES 15 mM. Five mL of versene-trypsin solution (BioLab, Jerusalem, Israel) are added to each flask, and the flasks are incubated at room temperature for 5 min to remove astrocytes. The flasks are shaken to dislodge the cells. The versene-trypsin solution is neutralized with 5 mL of S2. The cell suspension is collected and centrifuged at 100 g for 10 min. The supernatant is removed and the cells resuspended in S2. Cells are inoculated into 75-cm^2 flasks or plated in 35-mm dishes (Corning, Corning, NY, USA) and incubated until confluent at 37 °C 10% CO_2.

Poly-L-lysine Coating of Dishes

One mL of 10 µg/mL poly-L-lysine hydrobromide (Sigma, Rehovot, Israel) in 0.15 M boric acid, pH 8.4 (Merck, Germany), is added to 35-mm dishes. After 1 h, dishes are washed three times with sterile double-distilled water (sDDW) and kept at room temperature *(3)*.

Enriched Neuronal Cultures

The cells are prepared as described for rat cerebral cortical astrocytes cell cultures. After suspending the cells in S2, they are centrifuged at 100 g for 5 min and the supernatant is discarded. The cell pellet is resuspended in solution S3 containing 5% heat inactivated horse serum (Biological Industries), 0.1% gentamycin,

0.1% penicillin-streptomycin-nystatin, 1% N3 (defined medium components essential for neuronal development in culture) *(4)*, 15 µg/mL 5'-fluoro-2-deoxyuridine (FUDR, Sigma, Rehovot, Israel), and 3 µg/mL uridine (Sigma, Rehovot, Israel) in DMEM. Cells are counted in a hemocytometer, diluted in S3, and seeded in poly-L-lisine-coated 35-mm dishes (3×10^6 cells/dish). The medium is changed the next day to S3 without FUDR and uridine. Cells are allowed to grow for 1 week at 37 °C 10% CO_2 before experiments are performed.

Mixed Neuroglial Cultures

Neurons are prepared as described for enriched neuronal cultures. Three hundred cells/35-mm dish are seeded on 8-day-old astrocytes prepared as described for rat cerebral cortical astrocyte cell subculture. Cells are allowed to grow for 1 week at 37 °C 10% CO_2 before experiments are performed *(4)*.

Cell Lines

COS-7 cells (African green monkey kidney cells) are cultured in DMEM containing 10% heat inactivated fetal calf serum, 2 mM L-Glutamine, 100 units/mL penicillin, and 0.1 mg/mL streptomycin (Biological Industries) in 5% CO_2 at 37 °C (growth conditions) *(5)*. Every 5 days cells are split using trypsin-EDTA solution B (Biological Industries).

PC12 cells (Pheochromocytoma cells) are incubated under similar conditions to COS-7 except the medium is supplemented with 8% instead of 10% heat inactivated fetal calf serum and includes 8% donor horse serum *(5)*. In order to induce PC12 differentiation to neuronal-type cells, 1×10^6 cells are seeded on poly-L-lysine-coated 90-mm plates and incubated with 100 ng/mL of nerve growth factor (NGF) for 2 days *(3)*. The medium is changed and another batch of NGF (100 ng/mL) is added for 2 additional days, until the cells acquire a differentiated morphology.

HEK293 (human embryonic kidney cells) are cultured in DMEM containing 10% heat inactivated fetal calf serum, 2 mM L-Glutamine, 100 units/mL penicillin, and 0.1 mg/mL streptomycin (Biological Industries) in 5% CO_2 at 37 °C (growth conditions) *(5)*. Every 5 days cells are split using trypsin-EDTA solution A (Biological Industries).

BJ (human fibroblasts) are cultured in DMEM containing 20% heat inactivated fetal calf serum, 2 mM L-Glutamine, 100 units/mL penicillin, and 0.1 mg/mL streptomycin, 0.025% gentamycin sulphate solution (Biological Industries) in 5% CO_2 at 37 °C (growth conditions) *(5)*. Every 5 days cells are split using trypsin-EDTA solution A (Biological Industries).

MCF-7 (adenocarcinoma cells from human breasts) are cultured in DMEM containing 10% heat inactivated fetal calf serum, 2 mM L-Glutamine, 100 units/mL penicillin, and 0.1 mg/mL streptomycin (Biological Industries) in 5% CO_2 at 37 °C (growth conditions) *(5)*. Every 5 days cells are split using trypsin-EDTA solution A (Biological Industries).

OVCAR-3 (adenocarcinoma cells from human epithelial) are cultured, according to manufacturer's instructions, in RPMI 1640 medium containing 20% fetal bovine serum, 2 mM L-Glutamine, 1.5g/L sodium bicarbonate, 4.5 g/L glucose, 10 mM HEPES, 1.0 mM sodium pyruvate, and 0.01 mg/mL bovine insulin in 5% CO_2 at 37 °C (growth conditions). Every 2 to 3 days cells are split using trypsin-EDTA solution A (Biological Industries).

NIH3T3 (mouse fibroblasts) are cultured in DMEM containing 10% heat inactivated fetal calf serum, 2 mM L-Glutamine, 100 units/mL penicillin, and 0.1 mg/mL streptomycin (Biological Industries) in 5% CO_2 at 37 °C (growth conditions) *(5)*. Every 5 days cells are split using trypsin-EDTA solution B (Biological Industries).

HT29 (adenocarcinoma cells from human colon) are cultured in DMEM containing 10% heat inactivated fetal calf serum, 2 mM L-Glutamine, 100 units/mL penicillin, and 0.1 mg/mL streptomycin (Biological Industries) in 5% CO_2 at 37 °C (growth conditions) *(6)*. Every 5 days cells are split using versene-trypsin (BioLab).

SY5Y (neuroblastoma cells) are cultured in RPMI 1640 medium containing 20% fetal calf serum, 2 mM L-Glutamine, 1% sodium piruvat, 1% amino acid, 100 units/mL penicillin, and 0.1 mg/mL streptomycin (Biological Industries) in 5% CO_2 at 37 °C (growth conditions) *(7)*. Every 7 days cells are split using versene-trypsin (BioLab).

NCI-H727 (non-small-cell lung carcinoid) are cultured in DMEM containing 10% heat inactivated fetal calf serum, 2 mM L-Glutamine, 100 units/mL penicillin, and 0.1 mg/mL streptomycin (Biological Industries), 0.1% gentamycin sulphate solution in 5% CO_2 at 37 °C (growth conditions) *(8)*. Every 5 days cells are split using trypsin-EDTA solution A (Biological Industries).

CHO (Chinese hamster ovary) are cultured in DMEM containing 10% heat inactivated fetal calf serum, 2 mM L-Glutamine, 1% L-proline, 100 units/mL penicillin, and 0.1 mg/mL streptomycin (Biological Industries) in 5% CO_2 at 37 °C (growth conditions) *(9)*. Every 5 days cells are split using versene-trypsin (BioLab).

Summary and Practical Application in the Neuropeptide Field

Primary cell cultures and cell lines are being used in an ever-increasing range of applications and scientific disciplines. It is important that scientists starting out to use these cultures consider the desirability and relative benefits of the many *in vitro* cell culture systems now available. Primary cells might appear to be the models most directly relevant for replicating *in vitro* responses, but continuous cell lines have some compelling advantages for standardization and safety. Tissue culture techniques are developing rapidly to meet the demands of new approaches in areas such as cell therapy and tissue engineering. Developments in these areas will benefit from attention paid to the basic cell culture issues that are key to promoting the quality of research and safety of new cell therapeutic products. Over the years, our own work in the neuropeptide field relied heavily on cell culture techniques, primarily on cell lines. Such studies paved the path to an understanding of neuroimmunology *(5,10)* and allowed the identification of novel neuroprotective proteins that are acting downstream from the neuropeptide vasoactive intestinal peptide (VIP) *(11–13)*, the identification of specific lipophilic analogs of VIP *(14)*, and the characterization of mechanisms of action *(15,16)*.

Acknowledgments

We thank Dr. Douglas E. Brenneman for his excellent collaboration during many years and for his vital contribution in setting up the primary cell cultures described here. Professor Illana Gozes is the incumbent of The Lily and Avraham Gildor Chair for the Investigation of Growth Factors and the director of the Adams Super Center for Brain Studies, The Edersheim Levie-Gitter fMRI Institute, and the Dr. Diana and Zelman Elton (Elbaum) Laboratory for Molecular Neuroendocrinology at Tel Aviv University and serves as the chief scientific officer of Allon Therapeutics Inc., Vancouver, Canada.

References

1. Scheaffer, A.N. Proposed usage of animal tissue culture terms (revised 1978). Usage of vertebrate cell, tissue and organ culture terminology. *In Vitro*, 1979; **15**(9): 649–653.
2. Brenneman, D.E., et al. Protective peptides that are orally active and mechanistically nonchiral. *J. Pharmacol. Exp. Ther.*, 2004; **309**(3): 1190–1197.
3. Pilzer, I., and Gozes, I. A splice variant to PACAP receptor that is involved in spermatogenesis is expressed in astrocytes. *Ann. NY Acad. Sci.*, 2006; **1070**: 484–490.
4. Romijn, H.J., et al. Nerve outgrowth, synaptogenesis and bioelectric activity in fetal rat cerebral cortex tissue cultured in serum-free, chemically defined medium. *Brain Res.*, 1981; **254**(4): 583–589.
5. Brenneman, D.E., et al. Complex array of cytokines released by vasoactive intestinal peptide. *Neuropeptides*, 2003; **37**(2): 111–119.
6. Dangoor, D., et al. Novel extended and branched N-terminal analogs of VIP. *Regul. Pept.*, 2006; **137**(1–2): 42–49.
7. Offen, D., et al. Vasoactive intestinal peptide (VIP) prevents neurotoxicity in neuronal cultures: Relevance to neuroprotection in Parkinson's disease. *Brain Res.*, 2000; **854**(1–2): 257–262.
8. Davidson, A., Moody, T.W., and Gozes, I. Regulation of VIP gene expression in general. Human lung cancer cells in particular. *J. Mol. Neurosci.*, 1996; **76**(2): 99–110.
9. Moody, T.W., et al. (N-stearyl, norleucine17)VIP hybrid is a broad spectrum vasoactive intestinal peptide receptor antagonist. *J. Mol. Neurosci.*, 2002; **18**(1–2): 29–35.
10. Gozes, Y., et al. Conditioned media from activated lymphocytes maintain sympathetic neurons in culture. *Brain Res.*, 1982; **282**(1): 93–97.
11. Brenneman, D.E., and Gozes, I. A femtomolar-acting neuroprotective peptide. *J. Clin. Invest.*, 1996; **97**(10): 2299–2307.
12. Bassan, M., et al. Complete sequence of a novel protein containing a femtomolar-activity-dependent neuroprotective peptide. *J. Neurochem.*, 1999; **72**(3): 1283–1293.
13. Gozes, I., et al. NAP: Research and development of a peptide derived from activity-dependent neuroprotective protein (ADNP). *CNS Drug Rev.*, 2005; **11**(4): 353–368.
14. Gozes, I., et al. Neuroprotective strategy for Alzheimer disease: Intranasal administration of a fatty neuropeptide. *Proc. Natl. Acad. Sci. USA*, 1996; **93**(1): 427–432.
15. Visochek, L., et al. PolyADP-ribosylation is involved in neurotrophic activity. *J. Neurosci.*, 2005; **25**(32): 7420–7428.
16. Divinski, I., et al. Peptide neuroprotection through specific interaction with brain tubulin. *J. Neurochem.*, 2006; **98**(3): 973–984.

4

Transfection of DNA into Cells

Inbar Pilzer, Inna Divinski, and Illana Gozes

Key Words: transfection; cells; DNA.

Introduction

The novel technique of transfection is a key technique for the analysis of mammalian cells, opening new possibilities in gene therapy *(1)*. Although early reports of success with transfection techniques were met with skepticism, the procedure became well established. Together with the development of recombinant DNA techniques, transfection afforded the manipulation of mammalian cells. This technique allows transferring DNA into cells. In this short chapter, emphasis is given to neuropeptide genes and peptide receptors.

Several techniques are used to accomplish transfection and apply it in genetic research.

The Calcium Phosphate Method

The chemical transfection method calcium phosphate co-precipitation was first introduced by Graham and van der Eb *(2)*. This method includes slow mixing of buffered phosphate solution with a solution containing calcium chloride and DNA that produces a precipitate. This solution is layered onto the cultured cells and is taken up by endocytosis. The efficiency of DNA transfection with this method depends on the experimental conditions: calcium chloride concentration, DNA concentration, and pH of the transfection solution. The calcium phosphate method can be used for both transient and stable transfections.

From: *Neuromethods, Vol. 39: Neuropeptide Techniques*
Edited by: I. Gozes © Humana Press Inc., Totowa, NJ

DEAE Dextran

Another chemical method is the diethylaminoethyl (DEAE)–dextran method. DEAE-dextran and DNA are mixed, and then the solution is added to cells in a similar manner to the calcium phosphate method. Using adherent cells, this method produces reliable results in transient transfection experiments *(3)* in contrast to the low efficiency that is produced in stable transfection.

Since many cell types are difficult to transfect using chemical methods, these transfection methods are rarely used nowadays.

DNA–Liposome (or Lipid–DNA) Complex

The DNA–lipid complex is commonly used due to its high efficiency, simplicity, and versatility. Liposomes are complexes of synthetic lipids. Most procedures use cationic liposomes, so that the positively charged liposomes bind to the negatively charged phosphates of the DNA molecules *(4)*. Sufficient liposomes are used so that the complex has a surplus positive charge. This positive charge is attracted to the negative charge of the sialic acid residues on the surface of the cells. The DNA–lipid complex seems to penetrate the cell by endocytosis. Most DNA goes to lysosomes and is degraded, but a fraction continues on to enter the nucleus. The most popular transfection reagents include Lipofectamine (Invitrogen, Carlsbad, CA, USA), Fugene (Roch, Basel, Switzerland), and Polyfect (Qiagen, Valencia, CA, USA).

To perform transfection of cells in culture, the transfection cells are harvested a day before, resuspended with medium (according to manual instructions of the transfection kit), and seeded in 6-well plates or 24-well plates at a concentration of 6×10^5 or 2×10^5 cells/well, respectively. DNA and liposome reagent are diluted in serum-free media, since the presence of serum disrupts the liposome–DNA complex. The DNA and liposomes are combined for a short incubation time to allow for complex formation. The DNA–liposome complex is then incubated with the cells *(3)*. The cells are incubated for ~24–48 h at 37 °C until reaching 40–80% confluences.

The liposome transfection methods are usually 5- to 100-fold more efficient than the chemical methods described above for transient transfection and 3- to 20-fold more efficient in stable transfection.

Physical Methods

Alternative methods of transfection are physical methods. These methods provide a good substitute for difficult-to-transfect cells such as certain cell types, for instance, primary cell cultures, that are resistant to transfection by chemical and lipososme methods. These methods include microinjection, ballistics, and electroporation *(3,5,6)*.

Additional Methods

There are additional transfection methods, including cationic polymers *(7)*, fusogenic peptide *(8)*, and/or chloroquine to increase efficiency *(9)*.

Summary and Practical Application in the Neuropeptide Field

The transfection of DNA into mammalian cells in culture is a crucial research tool. In order to perform successful transfection, one must take into consideration the following parameters: (1) choosing the appropriate cell line for the research; (2) choosing the transfection method that suits the goals of the experiment; and (3) optimizing the reaction conditions for the transfection into the particular cell type. Examples for successful transfection related to neuropeptide research are the transfection of G-coupled protein receptors to ascertain peptide binding and function, as exemplified in our own work for vasoactive intestinal peptide (VIP). As cited in the abstract to our work *(10)*, VIP is known to provide neuroprotection. Three VIP receptors have been cloned: VPAC1, VPAC2, and PAC1. A specific splice variant of PAC1 in the third cytoplasmatic loop, hop2, was implicated in VIP-related neuroprotection. We aimed to clone the hop2 splice variant, examine its affinity to VIP, and investigate whether it mediates the VIP-related neuroprotective activity. The PAC1 cDNA was cloned from rat cerebral astrocytes. Using genetic manipulation, the hop2 splice variant was obtained, then inserted into an expression vector, and transfected into COS-7 cells that were used for binding assays. Results showed that VIP bound the cloned hop2 splice variant. Stearyl-neurotensin(6-11) VIP(7-28) (SNH), an antagonist for VIP, was also found to bind hop2. In addition, VIP protected COS-7 cells expressing hop2 from oxidative stress.

Parallel assays demonstrated that VIP increased cAMP accumulation in COS-7 cells expressing hop2. These results support the hypothesis that hop2 mediates the cytoprotective effects attributed to VIP and demonstrate the use of transfection methods in the study of neuropeptide function.

Acknowledgments

Professor Illana Gozes is the incumbent of The Lily and Avraham Gildor Chair for the Investigation of Growth Factors and the director of the Adams Super Center for Brain Studies, The Edersheim Levie-Gitter fMRI Institute, and the Dr. Diana and Zelman Elton (Elbaum) Laboratory for Molecular Neuroendocrinology at Tel Aviv University and serves as the chief scientific officer of Allon Therapeutics Inc., Vancouver, Canada.

References

1. Calos, M.P. The potential of extrachromosomal replicating vectors for gene therapy. *Trends Genet.*, 1996; **12**(11): 463–466.
2. Graham, F.L., and van der Eb, A.J. A new technique for the assay of infectivity of human adenovirus 5 DNA. *Virology*, 1973; **52**(2): 456–467.
3. Kingston, R.E. Introduction of DNA into mammalian cells. In *Current Protocols in Molecular Biology*, F.M. Ausubel, R. Brent, R.E. Kingston, et al., eds., 1997; pp. 9.0–9.7.
4. Gao, X., and Huang, L. Cationic liposome-mediated gene transfer. *Gene Ther.*, 1995; **2**(10): 710–722.
5. O'Brien, J.A., et al. Modifications to the hand-held gene gun: Improvements for *in vitro* biolistic transfection of organotypic neuronal tissue. *J. Neurosci. Meth.*,2001; **112**(1): 57–64.
6. Yang, N.S., et al. *In vivo* and *in vitro* gene transfer to mammalian somatic cells by particle bombardment. *Proc. Natl. Acad. Sci. USA*, 1990; **87**(24): 9568–9572.
7. Azzam, T., and Domb, A.J. Current developments in gene transfection agents. *Curr. Drug Deliv.*, **1**(2): 165–193.
8. Shimizu, H., et al. Genetic and expression analyses of the STOP (MAP6) gene in schizophrenia. *Schizophr. Res.*, 2006; **84**(2–3): 244–252.
9. Cheng, J., et al. Structure-function correlation of chloroquine and analogues as transgene expression enhancers in nonviral gene delivery. *J. Med. Chem.*, 2006; **49**(22): 6522–6531.
10. Pilzer, I., and Gozes, I. A splice variant to PACAP receptor that is involved in spermatogenesis is expressed in astrocytes. *Ann. NY Acad. Sci.*, 2006; **1070**: 484–490.

5

Transgenic Delivery and Detection of GFP in Neuropeptide Neurons

J.L. Holter, J.S. Davies, P.-S. Man, T. Wells, and D.A. Carter

Key Words: transgene; rat; fluorescent protein; immunocytochemistry; Western blot; somatostatin; pituitary; cortex; striatum.

Introduction

The first bioengineered autofluorescent protein, GFP (green fluorescent protein), was developed more than 10 years ago *(1)*. In recent years both this fluorescent protein and numerous variants (alternatively colored, destabilized, etc.; see *(2)*) have been widely adopted in experimental neuroscience *(3)*. When expressed in cultured cells and transgenic animals, GFP provides for direct (*in situ*) cellular visualization and thereby optical selection of individual cells for morphological, physiological, and molecular characterization. For some studies, generalized expression of GFP (e.g., actin gene promoter-driven; *(4)*) may be useful in facilitating the analysis of individual cell types. In general, however, the principal value of FP expression lies in spatial resolution, obtained through restriction of expression to a subpopulation of cells. In this case a more refined transgenic approach is required in which GFP is expressed under the control of a cell-specific gene promoter. An elegant example of such an approach permitted the first electrophysiological analysis of the scarce growth hormone releasing hormone neurons in the hypothalamus *(5)*. In many cell biological studies, GFP is expressed as a fusion molecule together with the cellular protein under study. This is a powerful transgenic approach because it permits direct tracking of the production

From: *Neuromethods, Vol. 39: Neuropeptide Techniques*
Edited by: I. Gozes © Humana Press Inc., Totowa, NJ

and cellular trafficking of endogenous protein. Unfortunately, this approach is not viable for functional studies of small neuropeptides that may lose functional conformation and receptor compatibility when fused to the 27-kDa GFP protein.

In the present chapter we present a method that provides for the co-expression and cellular sorting of GFP and a functional neuropeptide within a monocistronic transgene construct. Co-expression and sorting is important because it allows the investigator to monitor the functional extent of transgene expression, while biological outcomes are tested during overexpression of agonist or antagonist peptides.

We also present methods for GFP detection that have been developed during our studies of transgenic rodents. These methods are scattered throughout the primary literature; while similar methodologies are common to many laboratories, we have found that the selection of particular protocols and reagents is crucial in order to maximize the visualization and quantitation of GFP. Novel strains of transgenic rodents are now commonly received into laboratories for single experiments. It is important to immediately apply refined techniques in order to minimize the use of experimental animals. Finally, we consider aspects of GFP cell isolation. FP-transgenic rodent strains are now increasingly being used for cell sorting approaches in which fluorescent cells are isolated and used, for example, in microarray analysis *(6)*. In the present chapter, we present methods for brain dissociation that permit selection of individual fluorescent neurons.

Design of Transgene Constructs for Delivery of Neuropeptides and GFP

Transgene Construct Design

GFP can be expressed in neuropeptide neurons using a variety of different transgene designs. Generic examples of designs that have been applied in peptide biology and neuroscience are illustrated in Fig. 1. Design selection is dependent upon the individual study and is influenced by the availability of cell-specific promoters.

As noted above, the fusion protein design (Fig. 1, type 1) can be incompatible with neuropeptide activity and is therefore generally not tenable for functional studies. Nevertheless, this

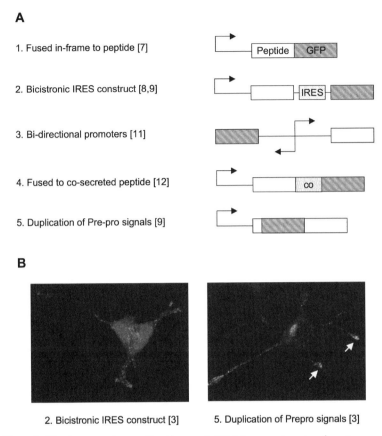

A

1. Fused in-frame to peptide [7]

2. Bicistronic IRES construct [8,9]

3. Bi-directional promoters [11]

4. Fused to co-secreted peptide [12]

5. Duplication of Pre-pro signals [9]

B

2. Bicistronic IRES construct [3]   5. Duplication of Prepro signals [3]

Fig. 1. Design and application of DNA constructs for transgenic expression of neuropeptides and GFP in cells. A. List of construct designs alongside schematic diagrams of generic constructs. Arrows denote promoter sequences. B. Differential expression of GFP in cultured AtT20 cells following transfection with either a somatostatin-IRES-GFP bicistronic construct, or alternatively a monocistronic somatostatin-GFP construct in which somatostatin pre-pro-sequence has been duplicated at the N-terminus of GFP. Note the diffuse expression of GFP throughout the cell obtained with the bicistronic IRES construct, compared with punctate expression obtained with the monocistronic construct, and also extensive co-localization of GFP with mature somatostatin peptide (prominent co-localization in two process terminal regions, indicated by white arrows). Images are fluorescence micrographs viewed with a confocal laser microscope (Leica TCS SP, Leica Microsystems Imaging Solutions Ltd, Cambridge, UK) using an x63 oil-immersion objective). IRES, internal ribosome entry site. co, co-secreted peptide. (Figure B is reproduced by permission from Davies, J.S., et al. *J. Mol. Endocrinol.*, 2004; **33**: 523–532. © Society for Endocrinology, 2004.)

design has been used to generate a GFP-tagged proatrial natri-uretic factor (ANF), which has provided valuable insights into the Ca^{2+}-stimulated selection of neuropeptide secretory granules *(7)*. Several alternative technologies are available for the co-expression of two separate protein/peptide products from a single transgene. The first (Fig. 1, type 2) utilizes an internal ribosome entry site (IRES) sequence to facilitate the translation of two products. This is a proven technology *(8,9)*, although bicistronic, which can result in unequal translation of the two products *(10)*. The second (Fig. 1, type 3) involves the use of two bidirectional promoters within a single construct. This approach has also been successfully used (and in combination with other transgenic technologies; see *(11)*), but commonly with strong viral (CMV, i.e., non-cell-specific) promoters. A third alternative approach (Fig. 1, type 4) is only applicable where the neuropeptide is associated with a co-secreted protein as with the neurophysins, which are encoded by both the oxytocin and vasopressin genes [see 12]. We have now developed an alternative approach for the co-expression and secretion of a biologically active neuropeptide and GFP from a monocistronic construct.

The "PEPS" Construct for Co-localization of Neuropeptides and GFP

The PEPS acronym stands for **p**re-**e**GFP-**p**rosomatostatin, as described in Davies et al. *(9)*. In this construct we have engineered the somatostatin and (enhanced) EGFP cDNA sequences into an open reading frame that can be expressed under the control of gene promoters of choice. However, GFP and somatostatin derived from this construct are not expressed (maturely) as a fusion protein, because the GFP is the positioned N-terminal of the somatostatin pre-pro-cleavage site. Hence, prosomatostatin is cleaved from GFP during processing, and a functional somato-statin peptide is derived. Furthermore, we have also duplicated somatostatin pre-prosequences N-teminal of GFP; consequently, GFP is processed to produce a product that bears N-terminal prose-quence and C-terminal pre-sequence. Therefore, this modified GFP product should contain signals that target GFP, like somatostatin, to the regulated secretory pathway (RSP).

We have tested the functionality of the PEPS construct in transiently transfected AtT20 cells (Fig. 1B) and compared this with a standard IRES construct (see above) in which GFP is expressed in an unmodified form. Fluorescence microscopic analysis of the transfected cells revealed a marked difference in GFP expression; only the PEPS construct generated spatially restricted GFP fluorescence that was localized to the same cellular compartments as mature somatostatin (Fig. 1B). As expected, the IRES construct successfully generated both mature somatostatin peptide and GFP, but the latter was abundant throughout the cells and did not concentrate at all into the granular (RSP-type) pattern that was observed for somatostatin. In further co-localization and immunoblot studies, we have confirmed that trafficking of the PEPS-derived GFP is consistent with concentration into secretory granules *(9)*. Accordingly, this novel construct design permits optical localization of transgene-derived neuropeptides without compromising their activity. Moreover, the monocistronic design should generate a 1:1 secretory ratio of GFP:neuropeptide, which could enable studies of neuropeptide secretion using GFP as a surrogate marker (see the upcoming section on analysis of GRP secretion). Finally, this approach should be considered generic and could easily be applied to other neuropeptide systems.

Detection of GFP in Transgenic Rodent Brain

Production and Genotyping of GFP Transgenic Rodents

Our application of GFP technology has focused mainly on the production and analysis of transgenic rat models that are generated using standard pronuclear microinjection procedures *(13)*. We use Southern blot analysis of genomic DNA to genotype founder animals and also to track the inheritance of individual transgene insertion events that may segregate at the F1 stage *(13)*. This analysis is vital because we have confirmed well-established findings of copy-number- and insertion-site-dependent transgene expression [see 13]. Following the establishment of stable lines of transgenics, genomic PCR can be used to genotype the offspring of matings between hemizygotic transgenics and wild types. The Sigma REDExtract-N-Amp kit (Sigma, St. Louis, MO, USA) is a rapid and robust genomic DNA extraction and PCR amplification system.

In order to confirm expression of GFP transgenes, we conducted an initial analysis at the RNA level. This can be necessary because, unlike some transgenic rodent lines where ubiquitous GFP expression permits direct visualization of fluorescing pups *(4)*, many lines exhibit discrete expression that may be confined, for example, within the brain. Furthermore, some GFP transgenic lines have failed to exhibit expression at the protein level *(14)* and so fluorescence is absent. Consequently, analysis conducted at the RNA level can serve to confirm (or not) that the transgene is adequately transcribed. We have used both Northern blot and RT-PCR procedures *(13)* to quantitate GFP-transgene RNA levels and, despite an absence of GFP protein in one transgenic model, have obtained valuable data on promoter function *(14,15)*.

Western Blot Analysis of GFP Protein

Western (immuno) blot can be required in the analysis of GFP-transgenic models: *(1)* to confirm transgene protein expression if levels of fluorescence are ambiguous, and *(2)* for the comparative analysis of GFP levels within whole tissues. For example, we have used Western blots to obtain a quantitative comparison of pituitary gland GFP levels in different transgenic rats (Fig. 2). Proteins were extracted from tissues using a standard whole cell extraction procedure *(16)* using buffer containing protease (Protease inhibitor cocktail, P8340, Sigma) and phosphatase inhibitors (Phosphatase inhibitor cocktail 1, P2850, Sigma), and stored at −70 °C. Western blots were performed using standard procedures that are available in most laboratories. For detection, we used a monoclonal anti-GFP antibody (BD Living Colors 8362-1; BD, Palo Alto, CA, USA) in combination with a standard secondary antiserum (HRP-linked donkey anti-mouse IgG; NA931, Amersham, Chalfont St. Giles, UK; 1:10,000) for 30 min. Protein bands were detected with a chemiluminescent detection system (ECL-Plus, Amersham). It should be noted that other commercially available anti-GFP sera did not perform as well using this protocol. Although GFP is readily detected using this protocol, Fig. 2 clearly shows that other, unidentified, protein bands are also detected when using this antiserum at a 1:10,000 dilution. Care must therefore be taken to resolve the GFP on a high-percentage polyacrylamide gel (e.g., 15%) in order to obtain accurate estimates of protein mass. In addition, we have used recombinant EGFP protein as a positive control (Fig. 2). Recombinant GFP is available from different

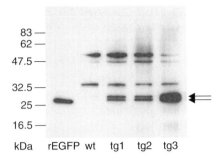

83 —
62 —
47.5 —
32.5 —
25 —
16.5 —

kDa   rEGFP   wt   tg1   tg2   tg3

Fig. 2. Western blot analysis of GFP proteins showing the use a monoclonal anti-GFP antibody (BD Living Colors 8362-1; BD, Palo Alto, CA, USA) to detect GFP protein in transgenic animal tissues. A control recombinant EGFP (rEGFP, 1 ng) is detected as a 27-kDa single band, whereas the d2EGFP derived from transgenic animal tissues (tg) is detected as a doublet (arrows) of approximately 28 and 30 kDa, or as a wide band in highly expressing tissues (tg3 sample). The greater mass of d2EGFP is due to the incorporation of destabilizing sequences derived from the mouse ornithine decarboxylase gene. Tissue extracts (50 mg) were resolved on a 15% denaturing polyacrylamide gel for 1.75 h at room temperature, transferred to a PVDF membrane (Hybond-P; Amersham Biosciences, Chalfont St. Giles, UK), and probed with the primary antibody (1:10000) for 45 min; then it was probed with a secondary antiserum (HRP-linked donkey anti-mouse IgG; NA931, Amersham, 1:10,000) for 30 min. Proteins bands were detected with a chemilumi-nescent detection system (ECL-Plus, Amersham) and X-ray autoradio-graphy (2 min exposure). Note the presence of larger molecular mass cross-reacting bands that are detected in both transgenic (tg) and wild-type (wt) tissue samples. Also note the presence of a faint 30-kDa band in the wt sample. Here, this band is detected at a significantly lower intensity compared with the d2EGFP bands but may give rise to false-positive data.

commercial sources, although it is expensive; therefore, investi-gators may wish to depend upon the specificity of expression in transgenic tissues in order to verify the presence of GFP. In this respect it should be noted that we have observed relatively low amounts of a cross-reacting (30 kDa) protein band that is (variably) detected in wild-type tissues (Fig. 2). This band can be excluded from interference in quantitative procedures by reducing film exposure times until the band is absent from wild-type samples.

Immunohistochemical Detection of GFP Protein

The use of immunohistochemistry to detect GFP protein may seem perverse given the autofluorescent properties of GFP, but this approach is widely used. GFP autofluorescence may be maintained for long periods in appropriately stored tissue samples (see next section), but the level of fluorescence becomes reduced compared with fresh tissue. Most importantly, immunohistochemistry provides a stable preparation that, depending upon the detection method (see the upcoming section on immunohistochemical procedures), can be stored for periods of months or years for retrospective analysis. Multiple labeling of tissue sections with GFP together with cellular antigens will be similarly maintained.

Tissue Preparation and Fixation

GFP protein can be detected by immunohistochemistry in tissues that have been prepared by a number of different protocols. Direct freezing of tissues prior to sectioning and brief paraformaldehyde fixation on microscope slides have worked effectively for EGFP *(12)*. A recent paper has evaluated different tissue-freezing protocols for GFP detection *(17)*. However, GFP will leak from unfixed cells; therefore, some form of immediate fixation procedure is often preferred. We have used 4% paraformaldehyde fixation for detection of destabilized EGFP (d2EGFP, BD) in rat tissues (Fig. 3). Both *in situ* fixation (via aortic perfusion, 200 mL 4% paraformaldehyde, followed by 90 min. postfixation, 4 °C, cryoprotection in 20% sucrose, overnight, 4 °C) and postfixation alone (immediately postdissection, 4% paraformaldehyde, 24 h, 4 °C, same cryoprotection) can be used, although in adult tissues we have found that the former is most effective in maintaining high levels of GFP immunoreactivity. Figure 3 shows a comparison of fixation methods for the adult rat brain cortex. It can be seen that there is a higher level of cellular GFP immunoreactivity in Fig. 3A (aortic perfusion), with maintenance of detectable immunoreactivity in long neuronal processes. In contrast, the detectable immunoreactivity in Fig. 3B (postfixation) is sparse, giving the appearance of "ghost cells." This contrast does appear to be due to differences in the efficiency of tissue penetration by fixative, because we have found that GFP immunoreactivity in neonatal rat brain tissue is very effectively maintained by postfixation. The antigenicity of other cellular proteins including

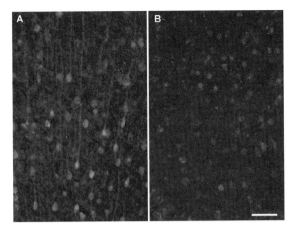

Fig. 3. Differential detection of GFP in transgenic rat brain sections following different tissue fixation protocols. Fluorescence micrographs of cortical neurons in rat brain sections following immunocytochemical detection of GFP protein (fluorophore, Alexa-488). **A**. Brain fixed *in situ* by intracardiac perfusion. **B**. Brain post-fixed following dissection. Note the lack of cellular detail in (cytoplasmic filling, neuronal processes) in (B). Bar = 50 mm. Images were viewed with an epifluorescence microscope (Leica DM-LB), and images were captured using a Leica DFC-300FX digital camera and Leica QWin software (V3).

nuclear transcription factors are generally very well maintained by either of these fixation protocols in adult tissues. With respect to GFP autofluorescence, the fixation method is also important because it has been shown that autofluorescence is reduced by paraformaldehyde fixation *(18)*.

Immunohistochemical Procedures

To detect GFP in transgenic rat tissues, we have used a standard fluorescence immunohistochemical procedure. Tissue sections are stored frozen (–70 °C) in black plastic staining troughs with inserted racks (L4163, Agar Scientific, Stanstead, Essex, UK). Both GFP immunoreactivity and autofluorescence are maintained for many months/years under these conditions. The main steps of the detection procedure are blocking (10% normal goat serum in 0.15% Triton X-100 phosphate buffered saline, 20 min,.room temperature), primary antibody incubation (see below, 1 h, room temperature), and fluorophore-labeled secondary antibody incubation (e.g., Alexa Fluor-488 goat, anti-rabbit IgG, Molecular

Probes/BD, 30 min, room temperature). Sections are mounted in Vectashield with DAPI (H-1200, Vector Laboratories, Burlingame, CA, USA) and stored in the same staining troughs at 4 °C. Fluorescent labeling is maintained for many months/years under these conditions.

We have found that two anti-GFP sera work effectively in this immunohistochemical procedure: rabbit anti-GFP (A11122, Molecular Probes/BD) and mouse anti-GFP (A11120, Molecular Probes/BD). The rabbit antiserum is superior, producing a brighter GFP signal in combination with a lower background. However, the mouse monoclonal antibody is useful for multiple labeling procedures where a "different-species" primary antiserum is required. In the context of multiple labeling, we have found that a GFP filter system (excitation filter: BP 470/40; dichromatic mirror: 500; suppression filter: BP 525/50) enhances the acuity of visualizing both Alexa Fluor-488 and GFP autofluorescence on our epifluorescence microscope. The signal-to-background quality of the image obtained with this filter set is considerably improved compared with a standard FITC filter system (excitation filter: BP 450/490; dichromatic mirror: 510; suppression filter: LP 515). This is particularly evident when multiple labeling is conducted with a "red" secondary fluorophore (e.g., Cy3) where the "green" image becomes contaminated with a red background. An extreme example of this phenomenon is shown in Figs. 4A–C, which show rat median eminence somatostatin detected with the Cy3 fluorophore. The high level of somatostatin immunoreactivity in this area of the rat brain is clearly detected with the FITC filter system (Fig. 4B), whereas no red fluorescence is detected at all when employing similar exposure times with the GFP filter set (Fig. 4C). Another methodological consideration is that GFP autofluorescence can be maintained for long periods when stored under the conditions described above; consequently, labeling other proteins with "green" fluorophores such as Alexa Fluor-488 can result in overlapping of the two fluorescent signals (see Fig. 4D).

Analysis of GFP Secretion

An unexpected facet of GFP technology is that this "foreign" protein can be efficiently processed and secreted by mammalian cells. This finding has demanded methods for quantifying secretion from transgenic tissues, e.g., endocrine organs *(19)*. We have

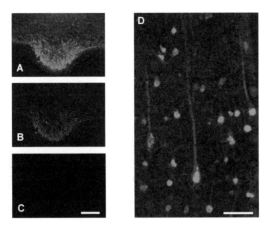

Fig. 4. Caveats to fluorescence detection. **A-C**: Detection of red fluorescence in rat brain sections using an FITC filter. Fluorescence micrographs of hypothalamic (median eminence) somatostatin nerve fibers viewed using different filter systems: **A.** "Red" (excitation filter: BP 515-560; dichromatic mirror: 580; suppression filter: LP 590) **B.** "FITC" (excitation filter: BP 450/490; dichromatic mirror: 510; suppression filter: LP 515) **C.** "GFP" (excitation filter: BP 470/40; dichromatic mirror: 500; suppression filter: BP 525/50). Note the detection of significant levels of red fluorescence using the FITC filter set but not the GFP filter set. Bar = 100 mm **D**. Maintenance of GFP fluorescence in transgenic rat brain sections following long-term storage. Fluorescence micrograph of cortical neurons following immunocytochemical detection of Egr-1 protein (bright [green] nuclear staining; fluorophore, Alexa Fluor-488). Note the presence of fluorescence in neuronal processes which is transgene-derived GFP remaining in cells following 6 months' storage of brain sections at –70 °C. Bar = 50 μm. Images were viewed with an epifluorescence microscope (Leica DM-LB), and images were captured using a Leica DFC-300FX digital camera and Leica QWin software (V3).

not employed these methods, but an efficient radioimmunoassay (sensitivity 10 pg) for EGFP immunoreactivity has been developed *(19)*. GFP fluorescence can be quantified by fluorometry, but a recent study has demonstrated the feasibility of adapting a real-time PCR thermal cycler for this purpose *(2)*.

Isolation of GFP Expressing Neurons

Current studies are exploiting one of the most powerful capabilities offered by GFP technology, namely the use of GFP autofluoresence to facilitate the isolation of specific subpopulations of cells *(6,20,21)*. Such cell populations can be used for a variety of purposes but perhaps most advantageously for microarray analysis. This capability has been made possible by other technological advances, most notably the improvement in RNA amplification methods [see 20]. GFP fluorescent cells can be sorted by FACS *(21)*, but for a small population of specific neurons in the brain, manual sorting using micropipettes has been successfully applied *(6)* (for a detailed protocol, see http://mouse.bio.brandeis.edu/).

We have used a commercially available enzymatic dissociation kit (NeuroCult® 05715, StemCell Technologies Inc., Vancouver, Canada) to obtain dissociated suspensions of rat brain cells (Fig. 5). First, 1-mm slices (rat brain matrix, World Precision Instruments, Sarasota, FL, USA) of transgenic rat brains are collected into NeuroCult® collection solution, and then microdissected brain regions are minced with a scalpel blade. The tissue is then dissociated by incubation in NeuroCult® enzymatic dissociation solution and subsequently through a process of resuspension and trituration. The single-cell suspension must then be appropriately diluted in resuspension solution so that a manageable population of cells can be viewed under the fluorescence microscope. We have used shallow glass depression slides for viewing and manipulating

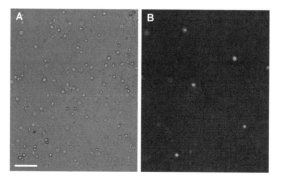

Fig. 5. Dissociation of GFP neurons from neonatal rat brain. Fluorescence micrographs of striatal neurons freshly dissociated from postnatal day 4 transgenic rats. A. Total cell content of camera field revealed with simultaneous bright-field and GFP filter system. B. Subpopulation of GFP autofluorescent neurons revealed with GFP filter system alone. Bar = 50 mµm. Images were viewed with an epifluorescence microscope (Leica DM-LB), and images were captured using a Leica DFC-300FX digital camera and Leica QWin software (V3).

small aliquots of the brain cell suspensions (Fig. 5). Individual fluorescent neurons can be collected using micropipettes and RNA extracted using an appropriate kit (e.g., PicoPure RNA Isolation Kit, Arcturus Bioscience Inc., Mountain View, CA, USA). We have found that this protocol is compatible with the use of a destabilized GFP (d2EGFP, BD), fluorescence being maintained in cell suspensions for several hours postsampling.

Concluding Remarks

GFP and other fluorescent proteins are an established part of the biologist's toolkit, but the efficient application of these tools does require the adoption of appropriate protocols and reagents. Details of generic protocols that have been described in this chapter can be obtained from laboratory manuals, for example, the Current Protocols series. Specific reagents do exhibit variable characteristics on a lot-by-lot basis, and investigators are urged to maintain empirical approaches to experimentation, particularly with respect to the use of commercially available antisera. It is hoped that the discussion of methods and approaches in this chapter will help to maximize the value of GFP transgenic animal models.

References

1. Chalfie, M., Tu, Y., Euskirchen, G, Ward, W.W., and Prasher, D.C. Green fluorescent protein as a marker for gene expression. *Science*, 1994; **263**: 802–805.
2. Utermark, J., and Karlovsky, P. Quantification of green fluorescent protein fluorescence using real-time PCR thermal cycler. *Biotechniques*, 2006; **41**: 152–154.
3. Spergel, D.J., Kruth, U., Shimshek, D.R., Sprengel, R., and Seeburg, P.H. Using reporter genes to label selected neuronal populations in transgenic mice for gene promoter, anatomical, and physiological studies. *Prog. Neurobiol.*, 2001; **63**: 673–686.
4. Okabe, M., Ikawa, M., Kominami, K., Nakanishi, T., and Nishimune, Y. "Green mice" as a source of ubiquitous green cells. *FEBS Lett.*, 1997; **407**: 313–319.
5. Balthasar, N., Mery, P.F., Magoulas, C.B., Mathers, K.E., Martin, A., Mollard, P., and Robinson, I.C. Growth hormone-releasing hormone (GHRH) neurons in GHRH-enhanced green fluorescent protein transgenic mice: A ventral hypothalamic network. *Endocrinology*, 2003; 144: 2728–2740.
6. Sugino, K., Hempel, C.M., Miller, M.N.,Hattox, A.M., Shapiro, P., Wu, C., Huang, Z.J., and Nelson, S.B. Molecular taxonomy of major neuronal classes in the adult mouse forebrain. *Nat. Neurosci.*, 2006; **9**: 99–107.

7. Burke, N.V., Han, W., Li, D., Takimoto, K., Watkins, S.C.,and Levitan, E.S. Neuronal peptide release is limited by secretory granule mobility. *Neuron*, 1997; **19**: 1095–1102.

8. Young, W.S., Iacangelo, A., Luo, X.Z., King, C., Duncan, K., and Ginns, E.I. Transgenic expression of green fluorescent protein in mouse oxytocin neurons. *J. Neuroendocrinol.*, 1999; **11**: 935–939.

9. Davies, J.S., Holter, J.L., Knight, D., Beaucourt, S.M., Murphy, D., Carter, D.A., and Wells, T. Manipulating sorting signals to generate co-expression of somatostatin and eGFP in the regulated secretory pathway from a monocistronic construct. *J. Mol. Endocrinol.*, 2004; **33**: 523–532.

10. Houdebine, L.M., and Attal, J. Internal ribosome entry sites (IRESs): Reality and use. *Transgenic Res.*, 1999; **8**: 157–177.

11. Krestel, H.E., Mayford, M., Seeburg, P.H., and Sprengel, R. A GFP-equipped bidirectional expression module well suited for monitoring tetracycline-regulated gene expression in mouse. *Nucleic Acids Res.*, 2001; **29**: E39.

12. Zhang, B.J., Kusano, K., Zerfas, P., Iacangelo, A.,Young, W.S., and Gainer, H. Targeting of green fluorescent protein to secretory granules in oxytocin magnocellular neurons and its secretion from neurohypophysial nerve terminals in transgenic mice. *Endocrinology*, 2002; **143**: 1036–1046.

13. Transgenesis techniques. In *Methods in Molecular Biology*, Vol. 18, D. Murphy and D.A. Carter, eds. Humana Press, Totowa, NJ, 1993.

14. Slade, J.P., Man, P.-S., Wells, T., and Carter, D.A. Stimulus-specific induction of an *egr-1* transgene in rat brain. *Neuroreport*, 2001; **13:** 671–674.

15. Man, P.-S., and Carter, D.A. Oestrogenic regulation of an *egr-1* transgene in rat anterior pituitary. J. Mol. Endocr., 2002; **30**: 187–193.

16. Smith, M., Burke, Z., Humphries, A., Wells, T., Klein, D., Carter, D., and Baler, R. Tissue-specific transgenic knockdown of Fos-related antigen 2 (Fra-2) expression mediated by dominant negative Fra-2. *Mol. Cell. Biol.*, 2001; **21**: 3704–3713.

17. Shariatmadari, R., Sipila, P.P., Huhtaniemi, I.T., and Poutanen, M. Improved technique for detection of enhanced green fluorescent protein in transgenic mice. *Biotechniques*, 2001; **30**: 1282–1285.

18. van den Pol, A.N., and Ghosh, P.K. Selective neuronal expression of green fluorescent protein with cytomegalovirus promoter reveals entire neuronal arbor in transgenic mice. *J. Neurosci.*, 1998; **18**: 10640–10651.

19. Magoulas, C., McGuinness, L., Balthasar, N., Carmignac, D.F., Sesay, A.K., Mathers, K.E., Christian, H., Candeil, L., Bonnefont, X., Mollard, P., and Robinson, I.C. A secreted fluorescent reporter targeted to pituitary growth hormone cells in transgenic mice. *Endocrinology*, 2000; **141**: 4681–4689.

20. Carter, D, Cellular transcriptomics—The next phase of endocrine expression profiling. *Trends Endocrinol. Metab.*, 2006; **17**: 192–198.

21. Lobo, M.K., Karsten, S.L., Gray, M., Geschwind, D.H., and Yang, X.W. FACS-array profiling of striatal projection neuron subtypes in juvenile and adult mouse brains. *Nat. Neurosci.*, 2006; **9**: 443–452.

6

Integrative Technologies for Neuropeptide Characterization

Sergey E. Ilyin and Carlos R. Plata-Salamán

Key Words: neuropeptide; HTS; siRNA; bioinformatics; TaqMan; Allegro™; RT-PCR; functional informatics; gene expression; target validation; biomarkers.

Microarray-based Platforms for Expression Analysis

Localization of expression provides an initial insight into the potential function of neuropeptides and related molecules. Expressional analysis is greatly facilitated by the recent advances in high-throughput technologies and in related bioinformatics tools to analyze, interpret, and manage the ever-increasing amount of data. Microarray technology offers a fairly straightforward approach for the analysis of gene expression. Microarrays are manufactured either by synthesis of probes directly on the chip surface or by spotting of presynthesized oligos or cDNAs on an appropriate chip surface *(1)*. The combination of photolithography and solid-phase synthesis is used to produce light-directed DNA synthesis on the surface of the Affymetrix chips (http://www.affymetrix.com, Santa Clara, CA, USA); light is directed to specific areas to produce a localized deprotection. Chemical reactions occur at the illuminated sites, and the cycle is repeated to generate 25-mer probes. NimbleGen (http://www.nimblegen.com/, Madison, WI, USA) also uses localized UV exposure to direct high-density DNA fabrication. Hundreds of thousands of individually addressable aluminum mirrors reflect the desired pattern of UV light and are controlled by computerized systems. Xeotron (http://www.xeotron.com, Houston, TX, USA) is also developing microarrays, using 3D microfluidic nanochamber chips and a

From: *Neuromethods, Vol. 39: Neuropeptide Techniques*
Edited by: I. Gozes © Humana Press Inc., Totowa, NJ

digital micromirror projector. Febit (http://www.febit.com/, Mannheim, Germany) uses a digital projector to direct DNA synthesis in 3D microchannel structures. The Febit device is available as a complete system, which includes DNA synthesis, microarray fabrication, sample introductions, hybridization, detection, and data analysis systems *(2)*.

There are two types of spotted microarrays: oligonucleotide and cDNA-based; the primary difference between these two is in the length of the sequence incorporated into the chip. Oligos are usually 20–35-nucleotide-long synthetic nucleic acids; cDNA arrays, on the other hand, incorporate much longer, generally 100–500-nucleotide sequences generated by PCR reactions. The CodeLink Bioarray platform from Amersham Biosciences (Chalfont St. Giles, UK) is produced by spotting prequalified 30-mers on a 3D gel matrix and the oligos are immobilized via covalent attachment to the matrix. The cDNA arrays are generated by spotting cDNAs on a glass slide. The design of these chips could be easily customized to individual researchers' needs. On the other hand, a drawback of the cDNA arrays is the difficulty of verifying the sequence identity of the clones spotted on the chip. Irrespective of the microarray type used in the analysis, a microarray experiment requires careful design and quality control of RNA samples and consists of several steps (Fig. 1). Before hybridization in a microarray, an RNA sample needs to be amplified and labeled. The type of label depends on the microarray platform used. CodeLink and Affymetrix GeneChip systems utilize a single-color

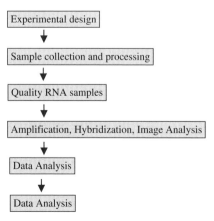

Fig. 1. General design of a gene expression experiment.

detection mode. Two-color platforms are competitive hybridizations in which one array is incubated with two different fluorescent samples. Different labeling strategies are discussed elsewhere *(3,4)*.

Integrative Tools for Microarray Data Analysis

Microarray image analysis is generally an automated process *(5)* that includes spot identification, segmentation, and intensity extraction. Data generated by image analysis require extensive quality control and normalization *(6,7)*. Minimum Information About Microarray Experiment (MIAME) is a set of rules aimed at making the microarray results reproducible (http://www.mged. org/Workgroups/MIAME/miame.html.). Normalization comprises a series of steps to adjust for the effects resulting from nonbiological differences between arrays. The experimental design also plays an important role in the success of an experiment. It is generally agreed that at least six biological replicates are required in a microarray experimental design to achieve statistically meaningful conclusions. The normalized microarray data undergo statistical analysis to identify genes of interest. A statistical microarray analysis package (SMA) is available at http://stat-www.berkeley.edu/users/terry/zarray/Software/smacode.html. SMA runs in an R statistical environment; R (www.r-project.org) is a widely used open-source object-oriented statistical software package. R provides a variety of statistical classification, clustering, classical statistical tests, and time-series analysis. R can easily be configured to generate publication quality and has a collection of tools for data analysis and display. An effective programming language, R includes defined functions, loops, and input and output facilities. Statistical analysis routines for analysis of microarray data have already been implemented in R. Several R-based packages have already been developed for this purpose, including Bioconductor (www.bioconductor.org) and DNAMR (www.rci.rutgers.edu/~cabrera/DNAMR/). R routines could be interfaced with other software, including Spotfire *(8)*. Spotfire DecisionSite/R is a guided application that provides a user-friendly interface for R routines. The data are then automatically sent to the R server, and the results of R's calculated statistics are returned seamlessly to DecisionSite. An example of hierarchal clustering performed by the Spotfire/R solution is presented in Fig. 2.

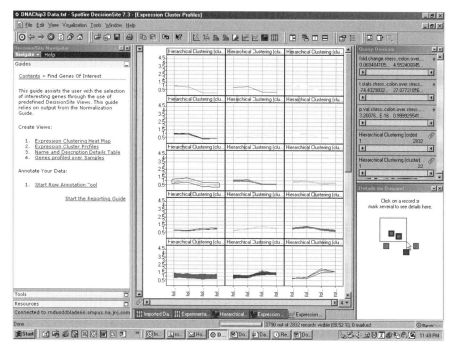

Fig. 2. An example of hierarchal clustering by Spotfire/R solution. Spotfire/R solution automates the analysis of microarray data. The example is courtesy of Stephen Prouty (Johnson & Johnson Pharmaceutical Research & Development, L.L.C.)

Emerging Technologies

Fixed-content arrays (which are mass-produced) can provide many data points; however, transcripts of interest may be missing. Moreover, the microarray infrastructure is expensive and requires significant informatics support. Microarray experiments also suffer from both false-positive and false-negative findings. The confirmation rate may vary depending on the microarray platform, type of statistical analysis, and method of confirmation. A significant opportunity exists when combining the throughput of a microarray with the sensitivity of PCR. For example, OpenArray™ Transcript Analysis (BioTrove) uses quantitative PCR to determine transcript copy numbers in samples of interest. OpenArray™ Microfluidics-based Transcript Analysis Platform (Fig. 3) consumes nanoliter-scale amount of reagents. Microscope slide-sized OpenArray™

Fig. 3. OpenArray™ Transcript Analysis System. The System performs 9,000 real-time PCR reactions in 2 h. (The image is courtesy of BioTrove Inc.)

plates (Fig. 4) contain over 3,000 through-holes, which accept 33-nanoliter reactions.

Validation of Microarray Data

Irrespective of the microarray platform used in the analysis, validation of the microarray data is crucial. The microarray data can be confirmed using several independent technologies. RNase protection could be a method of choice if real-time PCR equipment is not available. RNase protection assays are relatively easy to set up, and this method of gene expression analysis does not require any sophisticated or expensive equipment. The RNase protection assay methodology also offers good sensitivity and specificity as it combines a liquid hybridization assay with a proofreading mechanism (all single-stranded RNA is digested by RNAses). One of the advantages of this technology is the ability to multiplex (use of several different probes in the same reaction) and to detect several different protected bands on the same gel *(9,10)*. A limitation of this technology is a relatively low throughput, as probes have to be generated for each gene of interest and conditions of the assay have to be determined for each of the genes; however, probe generation can be optimized *(11)*. The RNase protection assay also offers increased sensitivity over hybridization-based methods for expression analysis (Northern

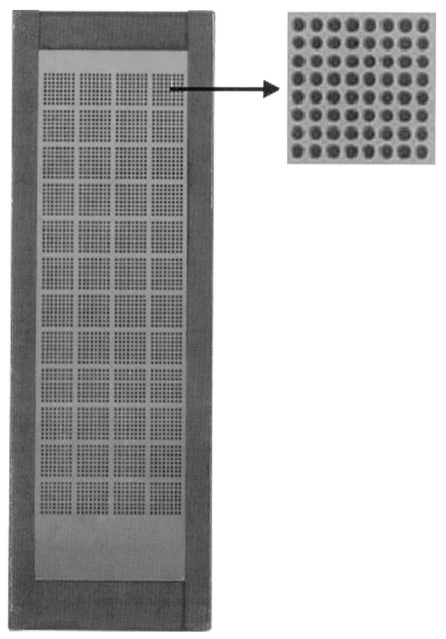

Fig. 4. BioTrove's microscope slide-sized OpenArray™ plates contain over 3,000 through-holes, which accept 33-nanoliter reactions.

and dot-blot), although detection is limited to relatively abundant mRNA molecules since no amplification steps are incorporated in the assay and only varying the length of exposure can regulate detection.

PCR-based methods of expressional analysis, on the other hand, offer almost unlimited sensitivity, as a single copy of DNA can be detected in the reaction tube. Several considerations for application of PCR to gene expression analysis are that (1) PCR conditions have to be uniform across samples, (2) the measurement needs to be done during the exponential phase of amplification, i.e., before the reaction reaches a plateau, and (3) the identity of a PCR product needs to be confirmed. Several modifications of qRT-PCR were designed and tested, including a competitive qPCR. Competitive qPCR effectively addresses the second issue, but it is difficult to develop and standardize. Furthermore, blotting of PCR products is often required to verify the identity of amplification products. Overall, the various considerations make a competitive PCR an unlikely method of choice for broad-based expressional analysis.

The TaqMan RT-PCR methodology was developed over the past decade. This approach offers a compelling choice for mRNA detection. This methodology combines real-time PCR with a liquid hybridizational assay. Figure 5 demonstrates and compares the sequence of steps in TaqMan and the RNase protection assay. In TaqMan RT-PCR, the product accumulation is measured in real time by the hybridization probe present in the reaction tube. Several factors that have contributed to TaqMan's popularity include fairly straightforward assay development; the assay

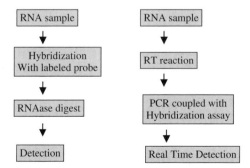

Fig. 5. Schematics of the RNase protection and TaqMan RT-PCR approaches.

conditions rarely require modifications; availability of pre-design reagents for a significant number (>30,000) of human genes; and suitable instrumentation delivering uniform well-to-well and plate-to-plate conditions. From the quantitation standpoint, there are two types of TaqMan RT-PCR: quantitative and semi-quantitative. From the experimental design point of view, there is little difference between these two assays. Both require incorporation of standards (or reference samples). In the case of quantitative RT-PCR, the investigator needs to know the copy of target in his reference tubes (which can be achieved by using a cDNA or even better *in vitro* transcribed mRNA to control for efficiency of translation). Once an RT-PCR run is completed, data are expressed in CT (threshold cycles or number of cycles it takes for a sample to cross the detection threshold). By comparing the CT of known versus unknown, or experimental, samples (Fig. 6), the exact copy

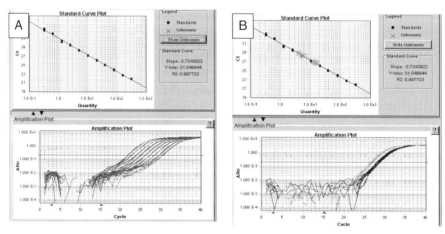

Fig. 6. TaqMan amplification plots showing amplification of reference samples and two groups of experimental samples. A. Reference samples ranging from 625 ng of RT material to 61 pg of RT material were analyzed in triplicates. Threshold cycle (CT) values were determined and a standard curve plot was constructed using the SDS 2.0 software package (Applied Biosystems, Foster City, CA, USA). A linear relationship between the amount of input material and CT values was observed. B. A standard curve plot was analyzed using SDS 2.0 to determine the amount of target transcripts in samples treated with control or target-specific siRNA. (Reproduced with permission from the *Journal of Biomolecular Screening* (**16**).)

number of target mRNA in experimental samples can be determined. Semi-quantitative RT-PCR is identical in design, except that serial dilution of an unknown sample is used as a reference. To minimize experimental artifacts, this reference sample needs to be generated under the exact same conditions as those of the experimental sample. Serial dilutions in this design help to determine the amplification efficiency. Even though an investigator may not know a target copy in the sample she/he is going to use as a reference, she/he may fairly safely assume that (if the target is expressed in this sample), by performing a 1-to-1 dilution of this sample, the copy number will be reduced twofold. Under ideal PCR conditions, this would result in a one-cycle difference between these samples, as the amount of product doubles with each PCR cycle. In the real setting, however, amplification efficiency may vary and could significantly alter the interpretation of experimental results. Serial dilution of a reference sample analyzed in parallel with experimental samples enables us to take the efficiency of amplification into account and accurately estimates the relative abundance of a target mRNA. In addition to reference samples, it is advisable to perform parallel analysis of a control gene (i.e., a gene that is not expected to change in the course of an experiment). This type of analysis helps to avoid potential problems, which may arise if the concentration of samples was estimated incorrectly. Other controls may include -RT (to control for contamination with genomic DNA) and negative control to verify potential contamination. The expressional profile obtained in the course of a microarray experiment and validated by TaqMan RT-PCR provides information about specific expression of neuropeptides and their receptors. Novel modifications of TaqMan technology (Fig. 7) make this assay especially useful *(6)*. Maley et al. *(6)* described a single-well TaqMan RT-PCR assay. In this procedure, the investigators use poly-A mRNA capture plates, and all steps—including RNA extraction, RT, and PCR—are performed in the same tube, thus significantly enhancing the throughput of the assay. This technology is useful for microarray data validation and also enables the use of signatures obtained in microarray experiments for various screening applications. The expression analysis via microarray and PCR may also demonstrate how the expression is affected by different conditions *in vivo*, for example, under hypoxia and anorexia/calorie-nutrient restriction.

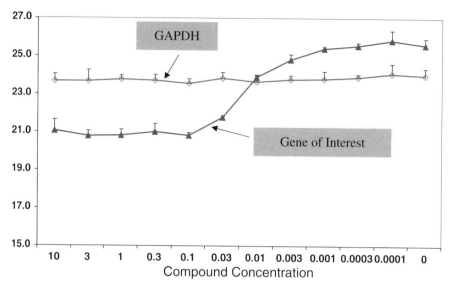

Fig. 7. Poly-A mRNA capture plates were utilized, and a single-tube RT-PCR assay for cell culture applications, including compound testing via gene induction measurement, was validated. In this example, all steps including RNA extraction, RT, and PCR are performed in the same tube, thus significantly enhancing the throughput of this method. The results were expressed as the threshold CT. Compound treatment did not alter levels of GAPDH, but produced significant induction of the gene of interest. Note the inverse relationship between CT and target copy number. See Ref. 6 for detailed descriptions. This example is courtesy of Daniel Horowitz (Johnson & Johnson Pharmaceutical Research & Development, L.L.C.).

Functional Characterization of Candidate Genes

An expressional profile may suggest a potential role, but direct functional studies are required to validate these observations. One approach for functional validation is to modulate the expression of a gene of interest *in vitro* or *in vivo*. *In vitro* experiments may be relevant for some of the neuropeptide functional characterization if the test system may be effectively reconstructed in cell cultures or other experimental *in vitro* systems of interest. Some of the potential functions and activities that could be studied *in vitro* may include neuroprotection, neuropeptide and cytokine production, and neurite outgrowth. A number of functional genomic tools may be used for these studies: siRNA (*12,13*), antisense, aptamers

(14), and different types of viral vectors. siRNA is a very selective and versatile tool, extensively reviewed elsewhere *(15)*. One of the advantages of siRNA relates to the mechanism of action and selectivity and specificity; siRNA experiments could be automated *(16)* and enable high-throughput functional testing (Figs. 8 and 9). A limitation often arises due to difficulties related to transfection of primary nervous system cultures and of most differentiated neuronal cell cultures. Problems with transfection could be generally overcome using a viral delivery system in which a molecule of interest is either overexpressed using a viral construct with an appropriate promoter, or the endogenous expression is suppressed using RNAi expressed by a viral construct. *In vivo*

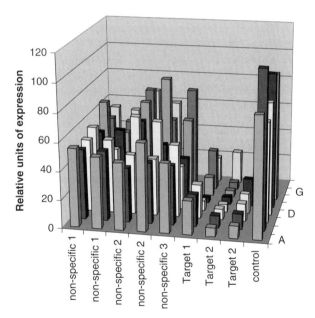

Fig. 8. Automated siRNA transfections were conducted in a 96-well format on the Allegro System. A target expression analysis was conducted on ABI PRISM® 7900 HT (Applied Biosystems, http://home. applied-biosystems.com). The Allegro High Throughput Screening system is amenable and adaptable to high-throughput siRNA strategies to assay for target validation. In this example, siRNA specifically and significantly downregulated appropriate transcripts, as detected by TaqMan quantitative RT-PCR. By designing functional assays to specific biological questions, siRNA can be used to validate targets in drug discovery. (Reproduced with permission from the *Journal of Biomolecular Screening* *(16)*.)

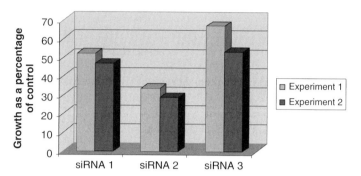

Fig. 9. Representative in a growth inhibition assay for three different gene-specific siRNAs run in triplicate in two different experiments. Each bar represents the growth of siRNA transfected cells as a percentage of control (control or scramble siRNA transfected cells = 100%). Growth inhibition was measured *in vitro*, using a Sulforhodamine B (SRB) assay as described earlier in Methods. Briefly, cells were plated in 96-well flat-bottom tissue culture plates at a density of 0.5×10^4 cells per well and allowed to grow for 24 h at 37 °C. All experimental measurements were performed in triplicate. Cells were transfected with siRNA and allowed to grow for 72 h. (Reproduced with permission from the *Journal of Biomolecular Screening (16)*.)

modulation of expression is often achieved by creating transgenic and/or knockout constructs. Information obtained in these models has provided very valuable insights on the function of neuropeptides, their receptors, and neuropeptide interactions with other endogenous chemical factors including neurotransmitters and cytokines. The direct delivery by application into the CNS of an agent that modulates expression has had some success *(17)*, but in many cases the data may be difficult to interpret. A question often arises from potential differential efficiency of delivery to glial, neuronal, and/or vascular components, time course and duration of an effect, as well as precise localization. With neuropeptides, often the method of choice is the direct administration of peptides into a brain region/area of interest *(18–20)*.

An example of an experimental paradigm using the direct brain administration of two classes of molecules, neuropeptide Y (NPY) and interleukin-1beta (IL-1beta), is shown in Fig. 10. This example demonstrates the robustness of the data that may be obtained in an *in vivo* model when studying functional interactions between endogenous ligands including neuropeptides. In this particular approach (which has also been integrated with cellular

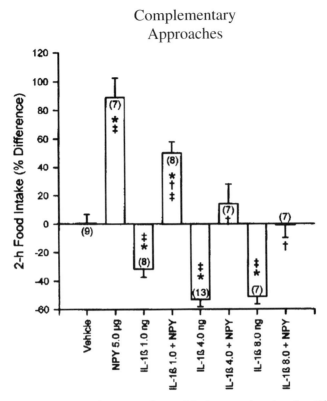

Fig. 10. Example of a functional peptide interaction *in vivo*. The intracerebroventricular microinfusion of NPY significantly increased 2-h food intake (by 89%) in rats; IL-1beta decreased 2-h food intake (32% with 1.0 ng/rat; 53% with 4.0 ng/rat; and 51% with 8.0 ng/rat). Vehicle infusion had no effect. NPY blocked the food intake suppression induced by IL-1beta. (Reproduced with permission from *Peptides* **(22)**.)

and molecular analyses), the brain administration of molecules is coupled with the microstructural analysis of behavioral modifications and the time course of action as well as with the analysis of brain regions and subregions (nuclei/areas) for expression of a molecule of interest (e.g., ligands, receptors). Overall, the relationship between expressional changes and behavioral modifications within an animal can be determined. In the example in Fig. 10, the intracerebroventricular microinfusion of NPY significantly increased 2-h food intake in rats, whereas IL-1beta decreased 2-h food intake **(22)**. This IL-1beta dose range includes doses that yield estimated pathophysiological concentrations in the cerebrospinal fluid. NPY blocked the anorexic effect induced by all doses of IL-1beta when both molecules were administered concomitantly. Central infusion of NPY was also able to induce feeding in IL-1beta-pretreated rats exhibiting marked anorexia. These results clearly show that NPY blocks and reverses the feeding suppression induced by estimated pathophysiological and pharmacological concentrations of IL-1beta. These results are mentioned within the context of a data set that shows the importance of studying peptide interactions in the brain that have demonstrable physiological, pathophysiological, and/or therapeutic relevance. Understanding the impact of endogenous peptide interactions on the expression level of molecules can aid in the mapping of integrative outputs of physiological events as well as in the characterization of pathophysiological cascades.

Opportunities in Platform Integration

Significant opportunities to advance our understanding of neuropeptide function exist in technology and platform integration.

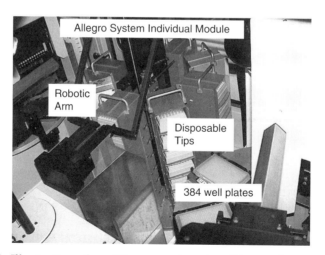

Fig. 11. Illustration of an Allegro workstation. The workstation consists of a host PC coupled to robotic stations for the purpose of sequentially processing a large number of samples residing in 96-well or 384-well microplates. The system accepts these samples as its input, processes them through precise combinations with the user-supplied reagents, and provides information and processed samples as its output. (Reproduced with permission from the *Journal of Combinatorial Chemistry and High Throughput Screening (23)*.)

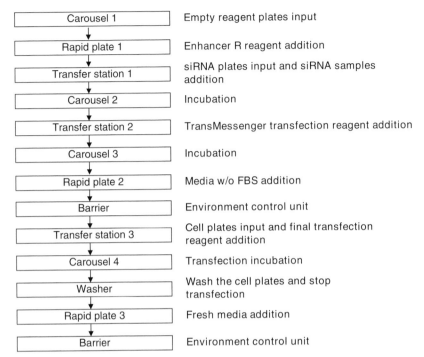

Carousel 1	Empty reagent plates input
Rapid plate 1	Enhancer R reagent addition
Transfer station 1	siRNA plates input and siRNA samples addition
Carousel 2	Incubation
Transfer station 2	TransMessenger transfection reagent addition
Carousel 3	Incubation
Rapid plate 2	Media w/o FBS addition
Barrier	Environment control unit
Transfer station 3	Cell plates input and final transfection reagent addition
Carousel 4	Transfection incubation
Washer	Wash the cell plates and stop transfection
Rapid plate 3	Fresh media addition
Barrier	Environment control unit

Fig. 12. The Allegro System (Zymark Corporation) consists of 13 modules in the following order: Carousel (CAR), Rapid Plate (RP), which is a 96-well pipettor, Transfer station (TR) consisting of an RP and two CARs (one containing disposable tips), CAR, TR, CAR, RP, Barrier (BAR), TR, CAR, Washer (WA), RP, BAR. The four modules enclosed by barrier units are environmentally controlled to maintain a temperature of 37 °C. (Reproduced with permission from the *Journal of Biomolecular Screening (16).*)

For example, "Functional Informatics" is a novel experimental paradigm based on the convergence and integration of bioinformatics and broad-based automation *(21)*. A fairly complex experiment could be split into a series of relatively simple steps, similarly to the approach used in automobile production (Figs. 11 and 12).

Recommended Experimental Protocols

RNA Extraction

The initial steps can be conducted according to the TRI reagent manufacturer's protocol (http://www.mrcgene.com; Molecular Research Center, Inc.).

Homogenization

Homogenize tissue/cells in TRI reagent (500 uL of reagent per 100 mg of tissue or 10^7 cells). Samples may be frozen at –85 °C.

RNA Extraction

1. Add molecular biology 0.1 mL of grade 1-bromo-3-chloropropane (BCP) per each 1 mL of TRI reagent used.
2. Vortex samples, and incubate for 12 min at room temperature.
3. Centrifuge at 13,000 rpm for 12 min at room temperature.

Precipitation

1. Carefully remove aqueous phase to a new tube.
2. Add 500 μL of isopropanol (per each mL of TRI reagent used).
3. Vortex samples, and incubate for 10 min at room temperature.
4. Centrifuge samples at 13,000 rpm for 10 min at room temperature.

RNA Wash

1. Decant liquid and wash pellet with 1 mL of 75% ethanol by inverting sample several times.
2. Centrifuge samples at 13,000 rpm for 5 min at room temperature.

Solubilization

1. Decant ethanol and air-dry pellet.
2. Dissolve pellet in 80 μL of RNase-free water.

DNAse Treatment

(Note: The authors tested a protocol for on-column DNase treatment and did not find the results satisfactory.)

This can be conducted according to the RNeasy Mini protocol for RNA Cleanup (Qiagen) and Rnase-Free DNase kit (Promega Cat. M6101).

1. Take 80 μL of sample and add 10 μL of 10X DNase reaction buffer (Promega Cat. # M6101).
2. Add 10 μL of DNAse, and incubate at 37 °C for 15 min.

Column Purification Using Qiagen (http://www1.qiagen.com) RNeasy Kit

1. Add 350 µL of RLT buffer containing beta-mercaptoethanol (10 µL of beta- mercaptoethanol + 1 mL of RLT buffer).
2. Add 250 µL of 100% ethanol to the sample, and vortex.
3. Transfer samples from Eppendorf tubes to columns.
4. Centrifuge at low speed until all the samples have been passed through the column. Start at lowest speed and gradually increase as sample is checked every 2 min or so.
5. Centrifuge at 13,000 rpm for 15 s to dry samples.
6. Apply column to a new collection tube.
7. Add 500 µL of RPE solution (make sure ethanol has been added). Centrifuge at 13,000 rpm for 15 s.
8. Apply column to a new collection tube.
9. Add 400 µL of RPE solution again. Centrifuge for 3 min at 13,000 rpm.
10. Apply column to a microcentrifuge tube with the lid cut off.
11. Add 30 µL of RNase-free water directly to the membrane (be sure not to touch the membrane with the pipette tip).
12. Incubate for 5 min at room temperature. Centrifuge at 13,000 rpm for 15 s.
13. Repeat steps 13 and 14.
14. When spin is over, keep the Eppendorf tube and discard the filter.
15. Check the optical density (OD) (260 and 280 wavelengths).
16. Store samples at −80 °C.

Detection of Optical Density

Parameters: 260 and 280 nm. Ratio 260/280 (range between 1.7 to 2.0).

Spectrophotometric conversion factor for RNA: 1 O.D. 260 nm corresponds to 40 µg/µL of RNA.

Reverse Transcription

A cDNA reaction was conducted according to a high-capacity cDNA archive kit from Applied Biosystems (Cat. 4322171).

1. Use up to 10 µg of RNA per reaction. Total volume: 40 µL, 0.25 µg/µL.

2. Total RT reaction: 80 µL, 40 µL of RNA sample, 40 µL of master mix (now it is 125 ng/µL).
3. Use template spreadsheet:
   For TaqMAN PCR, use 8 ng/µL. Total 40 ng/5 µL. 5 µL RT/reaction + 7 µL of master mix in 384 format. Total: 12 µL/well (384 format).

Calculation for References, Primers, and Probes

Primers' Preparation

720/(pmol concentration of primer) = x µL of water to add.
Final concentration is 900 nM for each primer and 250 nM for the probe.

Preparation of a PCR Reaction Mix

1. Calculate total volume (TV) you need; usually add 20% more.
2. Divide by 2 – volume of 2x master mix (Applied Biosystems) to add.
3. Divide TV by 200 or 20 (primers' volume). Divide by 200 if you design primers by yourself, and divide by 20 for custom primers.
4. Example: 100 reaction. 100 × 12 = 1200. 1200/2 = 600 (master mix), 1200/200 = 6 (primers), 5 (RT sample) × 100 (number of reactions) = 500. 1200 - 600 (master mix) – 6 (primers) – 500 (RT sample) = 94 µL of water.

Reference

RT sample is 125 ng/µL. 5 µL = 625 ng. We do not use this high concentration; we start with 312.5 ng/5 µL. Perform serial dilution:
1st: 312.5 ng of cDNA/5 µL
2nd: 156.25 ng of cDNA/5 µL
3rd: 78.125 ng of cDNA/5 µL
4th: 39.0625 ng of cDNA/5 µL
5th: 19.53125 ng of cDNA/5 µL
6th: 9.766 ng of cDNA/5 µL
7th: 4.88 ng of cDNA/5 µL
8th: 2.44 ng of cDNA/5 µL
9th: 1.22 ng of cDNA/5 µL
10th: 0.61 ng of cDNA/5 µL
11th: 0.305 ng of cDNA/5 µL

Example: Add 150 µL of water in each well. Add 150 µL of ref. cDNA to the first well (312.5 ng); mix well and transfer 150 to the following well; and so on.

Plate Preparation

1. Add 5 µL/well of RT sample (8 ng/µL).
2. Add 7 µL/well of PCR Reaction mix.
3. Seal the plates and spin.
4. Run experiment.

TaqMan Setup

(Set up before plate preparation.)

1. Open new document.
2. Choose detector (**Tools** – **detector manager**; select detector and then click: **copy to plate document** and **done**).
3. Select desired wells and put appropriate detector (on setup menu). If testing only one gene, select whole plate.
4. Insert plate to the machine (**instrument** – **open/close**; to close the door again: **instrument** – **open/close**).
5. On setup, remove first step.
6. Check your sample volume/well. For 384 format 12 µL.
7. Save document.
8. Start (instrument – start).

Primers' and Probe Preparation (Dilution)

●If you design primers and probe:

1 Divide Pmol number on 720 – amount of water you need to add to the tube with primer. Concentration will be 800x.
2 Probe comes ready to go at a 400x concentration.
3 Mix 1 volume of primer with 2 volumes of probe (1 + 1 + 2) – final concentration of the mixed assay will be 200x.
   Example: Primer1 Pmol = 40000: 40000/720.
If you order assay on demand (custom assay) – ready for use; 20x concentration.

Acknowledgments

The authors would like to thank Daniel Horowitz, Albert Pinhasov, and Stephen Prouty for their valuable comments and examples.

References

1. Hardiman, G. Microarray platforms—Comparisons and contrasts. *Pharmacogenomics*, 2004; **5**(5): 487–502.
2. Venkatasubbarao, S. Microarrays—Status and prospects. *Trends Biotechnol.*, 2004; **22**(12): 630–637.
3. Leung, Y.F., and Cavalieri, D. Fundamentals of cDNA microarray data analysis. *Trends Genet.*, 2003; **19**(11): 649–659.
4. Churchill, G.A. Fundamentals of experimental design for cDNA microarrays. *Nat. Genet.*, 2002; **32**(Suppl): 490–495.
5. Hariharan, R. The analysis of microarray data. *Pharmacogenomics*, 2003; **4**(4): 477–497.
6. Maley, D., Mei, J., Lu, H., Johnson, D.L., and Ilyin, S.E. Multiplexed RT-PCR for high-throughput screening applications. *Comb. Chem. High Throughput Screen.*, 2004; **7**(8): 727–732.
7. Forster, T., Roy, D., and Ghazal, P. Experiments using microarray technology: Limitations and standard operating procedures. *J. Endocrinol.*, 2003; **178**(2): 195–204.
8. Prouty, S., Nathan, D., Ledwith, J., Salisbury, M., Lyon, G., Messer, A., Amaratunga, D., Go, O.J.W., and Ilyin, S.E. Integrative tools for data analysis in pharmaceutical R&D. *PharmaGenomics*, 2004.
9. Ilyin, S.E., and Plata-Salaman, C.R. An approach to study molecular mechanisms involved in cytokine-induced anorexia. *J. Neurosci. Meth.*, 1996; **70**(1): 33–38.
10. Ilyin, S.E., Gayle, D., and Plata-Salaman, C.R. Modifications of RNase protection assay for neuroscience applications. *J. Neurosci. Meth.*, 1998; **84**(1–2): 139–141.
11. Ilyin, S.E., and Plata-Salaman, C.R. Probe generation by PCR coupled with ligation. *Nat. Biotechnol.*, 1999; **17**(6): 608–609.
12. Hammond, S.M., Bernstein, E., Beach, D., and Hannon, G.J. An RNA-directed nuclease mediates post-transcriptional gene silencing in Drosophila cells. *Nature*, 2000; **404**(6775): 293–296.
13. Elbashir, S.M., Harborth, J., Lendeckel, W., Yalcin, A., Weber, K., and Tuschl, T. Duplexes of 21-nucleotide RNAs mediate RNA interference in cultured mammalian cells. *Nature*, 2001; **411**(6836): 494–498.
14. Burgstaller, P., Girod, A., and Blind, M. Aptamers as tools for target prioritization and lead identification. *Drug Discov. Today*, 2002; **7**(24): 1221–1228.
15. Dykxhoorn, D.M., Novina, C.D., and Sharp, P.A.: Killing the messenger: Short RNAs that silence gene expression. *Nat. Rev. Mol. Cell. Biol.*, 2003; **4**(6): 457–467.
16. Xin, H., Bernal, A., Amato, F.A., Pinhasov, A., Kauffman, J., Brenneman, D.E., Derian, C.K., Andrade-Gordon, P., Plata-Salamán, C.R., and Ilyin, S.E. High-throughput siRNA-based functional target validation. *J. Biomol. Screen.*, 2004; **9**(4): 286–293.
17. Darrow, A.L., Conway, K.A., Vaidya, A.H., Rosenthal, D., Wildey, M.J., Minor, L., Itkin, Z., Kong, Y., Piesvaux, J., Qi, J., Mercken, M., Andrade-Gordon, P., Plata-Salaman, C., and Ilyin, S.E. Virus-based expression systems facilitate rapid target *in vivo* functionality validation and high-throughput screening. *J. Biomol. Screen.*, 2003; **8**(1): 65–71.
18. Plata-Salamán, C.R., Sonti, G., Borkoski, J.P., Wilson, C.D., and French-Mullen, J.M.B. Anorexia induced by chronic central administration of

cytokines at estimated pathophysiological concentrations. *Physiol. Behav.,* 1996; **60**(3): 867–875.

19. Sonti, G., Ilyin, S.E., and Plata-Salamán, C.R. Anorexia induced by cytokine interactions at pathophysiological concentrations. *Am. J. Physiol.,* 1996; **270** (6 Pt 2): R1394–1402.
20. Plata-Salamán, C.R. Leptin (OB protein), neuropeptide Y, and interleukin-1 interactions as interface mechanisms for the regulation of feeding in health and disease. *Nutrition,* 1996; **12**(10): 718–719.
21. Ilyin, S.E., Bernal, A., Horowitz, D., Derian, C.K., and Xin, H. Functional informatics: Convergence and integration of automation and bioinformatics. *Pharmacogenomics,* 2004; **5**(6): 721–730.
22. Sonti, G., Ilyin, S.E., and Plata-Salaman, C.R. Neuropeptide Y blocks and reverses interleukin-1 beta-induced anorexia in rats. *Peptides,* 1996; **17**(3): 517–520.
23. Pinhasov, A., Mei, J., Amaratunga, D., Amato, F.A., Lu, H., Kauffman, J., Xin, H., Brenneman, D.E., Johnson, D.L., Andrade-Gordon, P., and Ilyin, S.E. Gene expression analysis for high-throughput screening applications. *Comb. Chem. High Throughput Screen.,* 2004; **7**(2): 133–140.

.

7

Techniques in Neuropeptide Processing, Trafficking, and Secretion

Niamh X. Cawley, Tulin Yanik, Irina Arnaoutova, Hong Lou, Nimesh Patel, and Y. Peng Loh

Key Words: proneuropeptides; prohormones; regulated secretion; protocols; immunocytochemistry; pulse-chase labeling; Western blot; neurotrophins.

Introduction

Neuropeptides function as neurotransmitters and neuromodulators. They are synthesized as larger precursors at the rough endoplasmic reticulum (RER), trafficked to the trans-Golgi network (TGN), and sorted into granules of the regulated secretory pathway (RSP) for secretion in an activity-dependent manner. Polymorphisms found in human neuropeptide genes can lead to defects in trafficking and processing of the neuropeptide precursors, resulting in disease. Examples of mutations of human neuropeptide genes that have led to biosynthesis of precursors that were misrouted and only partially processed include insulin *(1)*, and cocaine and amphetamine-regulated transcript (CART) peptide *(2)*, giving rise to diabetes and obesity, respectively. A human valine to methionine mutation in the prodomain of brain-derived neurotrophic factor (BDNF) causes its inefficient sorting to the RSP and diminished activity-dependent secretion of BDNF from hippocampal neurons, resulting in memory deficits in these humans *(3)*. With the sequencing of the human genome, increasing numbers of polymorphisms in neuropeptide genes will

From: *Neuromethods, Vol. 39: Neuropeptide Techniques*
Edited by: I. Gozes © Humana Press Inc., Totowa, NJ

be identified. Studies on the trafficking, processing, and activity-dependent secretion of the mutant neuropeptide precursors will be useful in elucidating the molecular and cellular basis of diseases associated with the mutations. There are currently many paradigms and tools to study neuropeptide precursor trafficking, processing, and secretion, and these will be described in this chapter. These procedures are also applicable to studying the processing of other proteins such as neurotrophins.

Neuropeptide Biosynthesis and Processing

Three approaches to analyze neuropeptides processed from the precursor will be described. All three should be used preferably to fully identify the precursor and cleavage products.

Western Blotting with Different Antibodies

In this approach to analyze processing of neuropeptides, it is necessary to first identify a rich endogenous source of the neuropeptide of interest in a tissue or cell line. In the absence of such a source, the neuropeptide may be expressed in a foreign cell. Next, one should gather and generate as many antibodies as possible for the different parts of the precursor with a known amino acid sequence. Since processing occurs generally at pairs of basic residues, it is ideal to have specific antibodies to the peptide fragments that are flanked by the pairs of basic residues as well as to the N- and C-terminals of the precursor. Using these antibodies and the size of the immunopositive fragments observed in a Western blot of the tissue or cells, it will be possible to identify the various processed products generated from the precursor. This is an excellent first step to studying proneuropeptide processing. It should then be followed by pulse-chase labeling studies described below.

Pulse-Chase Labeling

The pulse-chase paradigm is a powerful means of tracking the synthesis and the dynamic processing of newly synthesized neuropeptide precursors. In this paradigm, cells are grown for a short period of time in a medium excluding the amino acid to be used for metabolic labeling. This medium is then replaced with the same medium containing a radioactive amino acid for a brief period of 10–30 min (the pulse). The radioactive labeling medium is then replaced by a medium containing up to 1 mM of the amino acid used as the label, so as to quickly dilute the radioactive pool. Cells are then incubated (the chase) in this medium for various times to allow the trafficking and processing of the newly synthesized precursor. The medium from each chase time point and the cell extract, harvested at the end of the experiment, are analyzed for processed products by immunoprecipitation with various antibodies (see the upcoming section on detection by immunoprecipitation and autoradiography). The appearance and disappearance of the different-sized processed products (intermediate forms) with time, coupled with the immuno-identity, allow the deciphering of the processing sequence.

Choice of Cells for Pulse-Chase Studies

Choosing the appropriate cell type for the pulse-chase studies is very important. In selecting the tissue or cells to study neuropeptide processing, it is best to choose a tissue, primary cultures, or a cell line that synthesizes the neuropeptide endogenously and in abundance. For example, to study pro-opiomelanocortin (POMC) processing, primary cultures of intermediate or anterior pituitary or the AtT20 cell line would be ideal since this proneuropeptide is abundant in these cells. If using an endogenous source proves too difficult, as in the case of generating primary neuronal cultures, and a cell line is not available, the next-best choice is to obtain a cell line with a regulated secretory pathway and then to transfect the cDNA of the neuropeptide precursor of interest into the cell line. Excellent model cell lines most commonly used are the AtT20 cells (ATCC, Manassas, VA, USA, Catalog #CRL-1795), which are a mouse corticotroph cell line; Neuro2A cells (N2A) (ATCC Catalog #CCL-131), a murine neuroblastoma cell line; INS-1 cells, a pancreatic β-cell line *(1)*; and GH3 and GH4C1 cells, somatomammotroph cell lines (ATCC Catalog #CCL-82.1 and CCL-82.2). PC12 cells (ATCC Catalog #CRL-1721), a rat pheochromocytoma cell line, do not contain the proprotein convertases PC1 and PC2, the processing enzymes, and should not be used for processing studies unless one is studying a protein that is known to be synthesized and processed by other enzymes in these cells. The advantage of using a cell line is that it can be commercially available and easy to maintain. However, the same cell line may differ slightly in different laboratories or even in the same laboratory, depending upon passage number and culture conditions. In general, cells should be low passage (<20). Another caveat is the cell line may not process the precursor exactly as it does *in vivo*, depending on the complement of prohormone convertases present in the cell line. It is generally a good approach to examine the processing in more than one cell line.

Transfection and Viral Infection Techniques

There are two commonly used ways of expressing an exogenous protein into cells in culture. One way is by transfection of a vector carrying a promoter and the encoding cDNA sequence of the protein of interest. Examples of such systems include pcDNA3.1 (Invitrogen, Carlsbad, CA, USA), pTARGET™ Mammalian Expression System (Promega, Madison, WI, USA), phCMV Expression Vectors (Gene Therapy Systems, Inc., San Diego, CA, USA), and BD Creator™ Gene Expression Systems (BD BioSciences, Franklin Lakes, NJ, USA). Transfection reagents commonly used to introduce the DNA into cells include Lipofectmine 2000 (Invitrogen) and SuperFect (Qiagen, Valencia, CA, USA). The accompanying procedures provided by the manufacturers are excellent starting points for expression determinations and will not be described here. However, for studies on regulated versus constitutive secretion, the optimal concentration of the DNA needs to be determined (see the upcoming section on overexpression).

Another way to express exogenous proteins is by infection with a virus-based DNA expression vector, e.g., adenoviral (available from BD BioSciences and Clontech), sindbis *(4)*, and Semliki Forest *(5)* virus-based DNA expression vectors. To introduce a protein to cells such as primary neurons, which have low transfection efficiency, using a virus-based DNA expression vector generally gives higher expression levels of the protein of interest due to increased numbers of transformed cells.

Protocol for Pulse-Chase Study

Metabolic labeling in the form of a pulse-chase protocol is a classical procedure and has been described in detail in *Current Protocols in Cell Biology* published by John Wiley and Sons, Inc., New York. We will, however, provide a brief description here with additional insights pertaining to the analysis of proneuropeptide and prohormone trafficking, processing and secretion. For clarity, all incubations are done in a humid tissue culture incubator (37 °C, 5% CO_2).

Cells to be labeled are rinsed gently and then incubated for 30 min with prewarmed DMEM medium that is devoid of the amino acid(s) to be used in the labeling but supplemented with dialyzed fetal bovine serum (dFBS). After this "starvation" period, the medium is replaced with new a starvation medium containing the appropriate radioactive amino acid so that protein synthesis can continue, and the radioactive amino acid is incorporated into newly synthesized proteins. The commonly used labeling compounds are [^{35}S]-methionine alone or a mixture of [^{35}S]-methionine and [^{35}S]-cysteine; however, other radioactively labeled amino acids can be used if they are abundant in the protein of interest, e.g., [^{3}H]-leucine in proinsulin or [^{14}C]-arginine in POMC. In choosing the label, it is important to be aware of the location of the radioactive amino acid in the prohormone. For example, pre-proinsulin has two methionine residues; however, these two methionines are in the leader sequence and hence would be removed immediately after synthesis. This would cause the remaining proinsulin to be unlabeled. Hence, for pulse-chase study of proinsulin, [^{35}S]-methionine labeling would not be appropriate. It is also important to obtain the highest specific activity for the radioactive label and use it in the procedure at a concentration of 100–200 µCi/mL in the labeling medium. The cells are incubated with this labeling medium for 10–30 min (pulse). The length of the pulse time depends on several parameters, for example, the abundance and rate of turnover of the protein of interest, the specific activity, and abundance of the radioactive amino acid present in the protein. As a control, one dish of cells is immediately harvested after the pulse (see below), representing the total amount of radioactivity that was incorporated during the pulse ($T = 0$).

To stop the labeling, the radioactive medium is removed and replaced with warmed complete DMEM supplemented with 10% fetal bovine serum (FBS), or alternatively, two volumes can be added straight to the labeling medium. This diluted labeling medium contains a sufficient amount of nonradioactive amino acids to rapidly quench the incorporation of the remaining pool of radioactive amino acids. If the cells may detach easily, a preferred method to stop the labeling is to add a concentrated solution of the nonradioactive amino acid to the labeling medium to give a final concentration of ~1 mM. The advantage of this is that there is minimal disturbance of the cells, and the volume of the chase medium is kept to a minimum, which might be important if antiserum for immunoprecipitations is limited.

For the purpose of studying secretion patterns, the chase media are collected at various time points and immediately treated with 1X Complete Inhibitor Cocktail (Roche, Indianapolis, IN, USA) to prevent protein degradation prior to analysis. All the media are centrifuged at 1,000 x g for 5 min in a microcentrifuge to remove floating cells and debris. Samples can be frozen until analyzed. These samples can be referred to as basal secretion samples containing proteins that were secreted from the cells in a nonregulated manner. To stimulate cells to secrete, the basal medium is replaced by fresh chase medium, typically a depolarizing medium containing 50 mM K$^+$ made iso-osmotic by reducing the concentration of NaCl, and cells are incubated for 10–30 min. Specific secretogogues may also be used, e.g., corticotrophin-releasing factor (CRF) to stimulate the release of adrenocorticotrophin (ACTH) from corticotrophs and glucose and glucagon-like peptide 1 (GLP-1) for the release of insulin from pancreatic β-cells. After this incubation, the stimulation medium is collected and processed as described above. The cells are rinsed twice with ice cold PBS and then incubated on ice for 20 min with 1% Triton-X-100/TNE (50 mM Tris, 150 mM NaCl, 2 mM EDTA, pH 7.2) buffer containing protease inhibitors. The cell homogenate is transferred to a 1.5-mL centrifuge tube, centrifuged at 15,000 x g for 10 min at 4 °C, and the supernatant is saved for analysis. The pellet is re-extracted with fresh TNE, and the combined soluble extract is stored at –80 °C until analyzed. The radioactive samples are generally analyzed by immunoprecipitation with specific antibodies.

Detection by Immunoprecipitation and Autoradiography

The principle of immunoprecipitation (IP) is to bind a specific antibody to a specific protein of interest in solution (in this case the protein of interest would be a mature neuropeptide and/or its precursor) and subsequently capture the antibody–antigen complex on a matrix that can be separated from the solution, usually Protein A-sepharose, or Protein G-sepharose beads, or agarose-coupled specific antibodies. Prior to IP, however, the sample should be cleared of proteins that bind nonspecifically to the sepharose beads by incubating it with Protein A- (or Protein G)-sepharose [30 µL of a 50% (v/v) slurry per 1 mL of sample] for 30 min at 4 °C with rotation. After a brief spin to sediment the beads, the supernatant can be retrieved and used for IP. We direct the readers again to *Current Protocols in Cell Biology* for a detailed protocol for IP and detection of the captured radioactive protein. Briefly, the antibody is added to the cleared sample usually at a dilution of 1:100 (but the dilution should be empirically determined), and the sample is incubated for 12–16 h at 4 °C with slow rotation. For an uncharacterized antibody, a serial dilution of the antibody should be carried out to optimize the IP procedure such that a second IP with a similar aliquot of antibody on the supernatant from the first IP should not yield further precipitated proteins of interest. It is important also to add fresh inhibitors to the sample prior to the IP in order to inhibit proteases that may be present in the antibody source. This is important if serum is used as the source of the antibody and is less important for purified antibodies. Protein A-sepharose beads are then added (10–50 µL per 1 mL of sample, depending on the concentration of antibody used and the binding capacity of the beads) and incubated for 30 min at 4 °C with rotation, after which the beads are collected by centrifugation and washed several times with appropriate buffers. The beads now containing antibody and antibody–antigen complexes are eluted by heating in SDS-PAGE sample buffer and analyzed by SDS-PAGE.

After gel electrophoresis, the gel is dried and exposed to a phosphorous screen (autoradiography), which is then recorded by a PhosphorImager (Amersham Biosciences, Uppsala, Sweden). To increase the sensitivity of the detection, especially for low-expression levels of neuropeptides, incubating the gel in a scintillation liquid (Ampify, Amersham) prior to drying will give a stronger signal for weak β-radiation emitters if the gel is subsequently developed by fluorography. Alternatively, the proteins in the gel can be transferred onto nitrocellulose or PVDF membrane and subsequently subjected to autoradiography. In the case of detecting pro- and mature brain-derived neurotrophic factor (BDNF), we found that transferring the proteins to membrane increased the detection level fivefold (Fig. 1) *(6)*. However, due to the differential transfer for proteins with various molecular sizes, such as processed peptide hormones and neuropeptides and their precursors, quantitative comparison of different size proteins

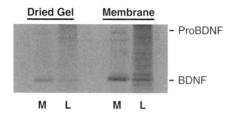

Fig. 1. Autoradiographic comparisons of signals of [35S]-labeled pro-BDNF and BDNF in a dried gel and on a transferred membrane. AtT20 cells were transfected with proBDNF, pulse-labeled with [35S]-Met/Cys, and then chased for 1 h. The medium (M) and cell lysates (L) were immunoprecipitated with BDNF antibody, and the eluted protein subjected to SDS-PAGE in duplicate. After electrophoresis, half of the gel was dried onto filter paper for 2 h, and the other half of the gel was transferred onto nitrocellulose membrane (0.45-µm pore size). Both were read on the PhosphoImager. The signal of both proBDNF and BDNF in the membrane (right panel) is five times stronger than that of the dried gel (left panel).

within a lane may not always be accurate unless equivalent levels of recovery onto the membrane can be verified. It is also important to be aware that not all peptides transfer equivalently nor do they bind identically to the membrane, and as such the correct method for detection of the specific protein of interest should be established, as subtle changes in procedures may have dramatic effects on the results. See also the section on steady-state analysis (Fig. 4); for quantification of radioactive bands, see the section under pulse-chase labeling analysis.

SELDI Protein Chip Technology

New methods for mass spectrometry have been developed to study neuropeptides (or small peptides) with greater sensitivity and accuracy. One of these methods is Protein Chip technology. Protein Chip technology (protein microarrays) uses surface-enhanced laser desorption/ionization time-of-flight (SELDI-TOF) mass spectrometry to detect peptides *(7,8)*. This method requires the Ciphergen ProteinChip assay system (Ciphergen Biosystems, Fremont, CA, USA), which is costly but, if available, provides a very sensitive assay needing only very small amounts of tissue, cell extract, cell culture medium, or plasma, which could be analyzed in a very short time. It is particularly useful in analyzing processed neuropeptide products in human plasma from normal versus disease states; few systems can handle plasma directly without some prepurification steps prior to analysis. The system can also be used for peptide sequencing and protein identification.

The ProteinChip System

The ProteinChip system performs selective protein extraction from biological complexes and retention on chromatographic chip surfaces followed by laser desorption/ionization mass spectrometry. In this procedure, proteins are first bound to the chromatographic surface of the chip and the unbound proteins are washed away. The adhered proteins/peptides are then treated with an acid matrix, consisting of sinapinic acid (SPA) or α-cyno-4-hydroxy cinnamic acid (CHCA) for proteins and peptides, respectively, to enhance ionization, and then air-dried onto the surface. The matrix solution is prepared in 50% acetonitrile water containing 1% trifluoroacetic acid. Captured protein samples on the chip surfaces, aligned as a row of eight spots in the array, are inserted into the ProteinChip reader, consisting of a vacuum chamber and where a laser beam blasts off to activate the desorption/ionization process, liberating gaseous protein ions from the spots. The ionized proteins fly down the vacuum tube toward an oppositely charged electrode. The mass to charge (m/z) value of each ion is the time it takes for the launched ion to reach the electrode; small ions travel faster. Therefore, the spectrum provides a time-of-flight signature of ions ordered by the size of the protein/peptide. The individual proteins/peptides are displayed as unique peaks based on their molecular mass using the computer software provided by the manufacturer and shown schematically in Fig. 2. Current mass spectra platforms have sensitivity in the femtomolar range and may become even more sensitive in the future (*9*).

ProteinChips

The availability of several chromatographic surfaces of ProteinChip® arrays enables binding of a large variety of proteins and peptides for protein profiling and peptide mapping applications. The most common ProteinChip arrays used for protein/peptide analysis are normal-phase (NP20) chromatography with silicate functionality or reverse-phase chromatography with C16 functionality (H4) or C6 to C12 functionality

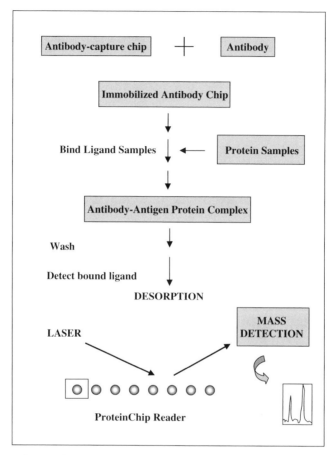

Fig. 2. General outline of the antibody capturing ProteinChip array protocol. Protein G is covalently attached to the chip surface. The antibody is incubated on the chip and is bound to the protein G. The sample is then incubated on the chip where the antibody can capture it. The chip is coated with an energy-absorbing matrix and the proteins analyzed by SELDI-TOF.

(H50). Additionally, antibody-capturing chips (PG20, for antigen-antibody reactions) are very useful for detecting the processing of larger precursors to their intermediate or active forms.

Normal-Phase ProteinChip Array (NP20)

These arrays are mostly for general protein-binding purposes and recommended for the binding of hydrophilic proteins/peptides. Active spots contain silicon oxide that allows proteins to bind via serine, threonine, or lysine residues. This type of array is useful for quick screening of proteins/peptides in a biological sample to verify the presence or absence of a molecule. For example, to determine the extent of processing of the proneuropeptide, POMC, in the intermediate pituitary of wild-type versus carboxypeptidase E (CPE) knockout (KO) *(10)* mice, the POMC processed products present in the profiles of the two tissues can easily be compared.

Hydrophobic Surface ProteinChip Array (H4/H50)

H4 and H50 chips are used for capturing proteins through their hydrophobic interactions. Active spots contain chains of up to 16 methylene groups that bind proteins through reverse-phase chemistry. It is mostly for rapid protein/peptide analysis and binds proteins/peptides abundant in alanine, valine, leucine, isoleucine, phenylalanine, tryptopan, or tyrosine. Protein samples can be directly applied to the ProteinChip spots, or the spots can be pretreated with different reagents such as salt or detergent to increase the hydrophobicity of the sample for binding. Prior to coupling protein/peptide sample to the spot, a hydrophobic pen should be used to circle the spot for sample containment within the spot.

Recently, the SELDI-TOF MS approach using H4 ProteinChips has been successfully applied to determine the amino acid sequence of peptides using carboxypeptidase Y (CPY) "on chip" *(7,11,12)*. Cool and Hardiman *(11)* used H4 arrays to identify the amino acids at the C-terminus of ACTH from dissected pituitary samples. Only 1 µg of the total pituitary protein was required. The tissue was extracted with 0.1 M HCl, applied onto the chip, and dried, and CPY was added. After various times, CHCA was added to enhance ionization of the molecules, and the chip was then scanned in the ProteinChip reader. The precise C-terminal amino acid sequence of ACTH was determined "on-chip." This approach is valuable for determining the C-terminal residues of cleaved products and hence the cleavage site within the precursor, provided the cleaved peptide is not located at the very C-terminus of the precursor. However, this approach can only be successful if the cleaved product is present in abundance in the tissue as in the case of ACTH in the pituitary.

Antibody Capture Chips (PG20)

Antibody capture chips are very useful for identifying processed neuropeptides *(12)*. PG20 ProteinChips arrays are preactivated surface (PS20) chips that are precoupled with recombinant protein

G, which can be used for immunoassay-based protein/peptide detection (Fig. 2). Any antibody capable of binding to protein G can be used in these arrays; however, it is important to use a polyclonal antibody that recognizes the precursor and processed products so that cleaved peptides from a larger protein can be detected. It is also important to determine the integrity of the antibody by analyzing it on the normal-phase chip (NP20) prior to the antibody capture procedure. Then the antibody can be added to the spots of the arrays directly and incubated for 1–4 h at room temperature (or overnight at 4 °C) in a humid chamber. If the antibody is in a solution containing sodium azide as a preservative, it is necessary to remove it prior to coupling, since azide will interfere with antibody coupling. Unbound antibodies are generally washed with PBS, but a detergent can also be added to the PBS solution in order to increase the stringency of the wash. Both 0.05% TritonX-100 and Tween-20 can be used. After washing, protein samples are added to individual spots on the chips and incubated for 1–4 h (or overnight at 4 °C) in a humid chamber. Following the incubation, the chips are washed with PBS or the above washing buffers and rinsed with distilled water quickly. After the spots are air-dried, CHCA (20%) is added to each spot on the ProteinChip and data acquisition is performed by the ProteinChip Reader.

An example of this application is shown in Fig. 3. Processing of CART peptide, an anorectic neuropeptide *(13)*, was analyzed using the antibody capture chip, in CPE KO mice *(10)*. Sera from wild-type and knockout (KO) animals (2 µL of 10X diluted serum) were bound onto the chips that were coupled with anti-CART (55–102). Figure 3 shows that processed forms of CART, CART I, and CART II (active forms) were detected in sera from wild-type but not in CPE KO animals. Serum from KO animals contained a partially processed intermediate form of CART. This approach is useful for identifying the presence of biologically active forms of the peptide in serum and tissue, and hence the processing pattern of the precursor *in vivo*.

Peptide Mapping

SELDI-TOF can also be used to identify peptides in biological or crude samples. By peptide-mapping the processed products, it will be possible to identify from which part of the precursor the peptides are derived. However, several purification steps may be

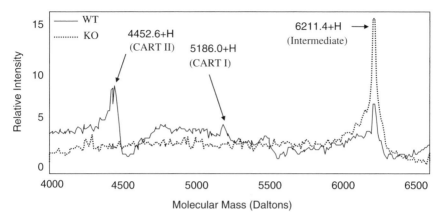

Fig. 3. Analysis of cocaine and amphetamine-regulated transcript (CART) in serum of CPE knockout (KO) mice using antibody capture ProteinChip arrays. Anti-CART(55-102) was coupled to protein G-coated spots on the chip array. Diluted sera from WT and CPE KO animals were then incubated on these spots; after washing, the peptides were analyzed by SELDI-TOF. CART I and II are processed active forms of proCART that were detected in the serum of WT animals (solid line). In contrast, only intermediate CART (unprocessed) was detected in the serum of CPE KO animals. No active forms were detected (dashed line).

necessary to obtain the pure peptide of interest for identification *(8)*. After the purified protein/peptide is obtained, it is incubated with trypsin for digestion and then applied on the protein microarray. Matrix is added for ionization, and mass analysis is performed by SELDI. Spectra should be internally mass-calibrated using trypsin autolysis for reference *(14)*. In order to identify the proteins, peptide masses are retrieved from the mass spectra and searched against protein databases to identify the protein fragment *(8,14)*. Identification of the neuropeptide fragments in the tissue using peptide mapping will allow deciphering the processing pattern of the precursor, whose sequence is known.

Neuropeptide Trafficking and Secretion

The following section deals with issues involved in the analysis of neuropeptide trafficking and secretion. We try to highlight potential problems that occur and address issues that may not be obvious but are important.

Steady-State Analysis

Analysis under steady-state conditions is the most common procedure used to study the secretion and processing of prohormones and proneuropeptides. Cells expressing the proteins of interest are rinsed twice with prewarmed basal medium (DMEM/0.01% BSA) and then incubated for several 30-min periods in order for the cells to become equilibrated with the new environment. These 30-min incubates, representing basal (nonregulated) constitutive secretion under steady-state conditions, are collected and analyzed. After incubation in basal medium, the cells are then incubated in a stimulating medium (for endocrine cells, we use DMEM/0.01% BSA supplemented with 50 mM KCl and 2 mM $BaCl_2$ and made isotonic by reducing the concentration of NaCl). Since depolarization occurs rapidly, the initial secretory response represents the secretion of mature granules in the readily releasable pool of granules. We therefore collect the stimulating medium after 10 min. Incubating the cells for longer times results in the mobilization of reserve pools of granules and indeed the trafficking and secretion of new granules formed at the TGN. Prior to analysis or storage at –20 °C, the samples are centrifuged at 1,000 x g for 5 min in a microcentrifuge in order to remove floating cells and cell debris. After collection of the stimulating medium, the cells are rinsed twice with ice cold PBS and harvested with ice cold lysis buffer (M-per Mammalian Extraction Buffer, Pierce, or 1% Triton X-100/TNE buffer) with fresh inhibitors added. The cells are centrifuged at 15,000 x g for 10 min and the soluble extract saved for analysis.

Secretion experiments can be performed using complete culture medium if the analysis can be carried out in the presence of 10% heat-inactivated FBS, i.e., by radio-immunoassay (RIA) or IP. Serum proteins provide an efficient means of prevention against degradation of the (pro) neuropeptides by proteases, as well as being a good blocker to prevent absorption of the protein of interest to the plastic found in the dishes and tubes where the samples are stored. In cases where some antibodies do not work for RIA or IP, immunoblotting is widely used. The limitations are that serum proteins would saturate the blot and give an unacceptable background. To circumvent this problem, BSA can be added to the secretion medium to a concentration of 0.01% in place of the 10% FBS. This amount of BSA is tolerated well on Western blots and is sufficient to provide stabilizing and protective functions for the secreted proteins that are to be analyzed. We use DMEM as a basic secretion medium because it contains all the necessary components for normal cellular metabolism (see the section on buffer systems). In many cases the level of the secreted protein may be low and require concentration. If the protein of interest is greater than 10 kDa in mass, the presence of the 0.01% BSA acts as a carrier when precipitated by 10% trichloroacetic acid (TCA). Briefly, 50 µL of 100% TCA are added to 450 µL of the secretion medium; after mixing well, the solution is allowed to precipitate on ice for 30 min. The samples are centrifuged for 20 min at 15,000 x g in a microcentrifuge and the supernatant discarded. The sample is then incubated in 1 mL of cold 100% acetone for 10 min on ice, followed by centrifugation for 10 min at 15,000 x g. The supernatant is discarded carefully, and the pellet is allowed to air-dry in a chemical fume hood. The pellet can then be reconstituted in 45 µL of 1X SDS-PAGE sample buffer and prepared for Western blot analysis. Recovery efficiency of the sample can be ascertained by a comparison in the recovery of the BSA as a marker for the consistency of the procedure. While proteins/peptides smaller than 10 kDa are not efficiently precipitated with 10% TCA, the presence of a carrier protein such as BSA will allow some recovery, such that its relative level in different samples can be assessed, rather than its amount relative to its precursor or intermediate within the same sample.

For the analysis of the processed peptide hormones and neuropeptides, Western blot has been used successfully for peptides such as ACTH, insulin, and CART; however, we have found that the choice of immunoblotting procedures (gel type, buffer system, transferring system, and membrane) can have a profound effect on the final result. An example of this is demonstrated in Fig. 4, where medium and lysate from a secretion experiment performed on AtT20 cells were analyzed by Western blot. Two different procedures were used, as described in the figure. The results show that ACTH is detected easily on the PVDF membrane and not on the nitrocellulose membrane, while POMC and its intermediate are detected very well on the nitrocellulose rather than the PVDF membrane (Fig. 4). Although these two procedures are obviously different, the extent to which the results differed was unexpected. Consequently, it is very important to establish the correct method for Western blot analysis of the specific protein of

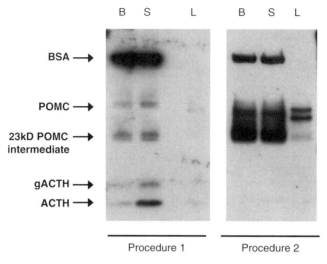

Fig. 4. Comparison of two methodologies for Western blot analysis of POMC related proteins. A secretion assay was performed on AtT20 cells, and the basal and stimulated medium, as well as the cell lysates, were analyzed in parallel by two procedures. One set was separated by a 16% Tris/Glycine PAGE gel and transferred (50 mA, 1 h) by a semi-dry immunoblotting apparatus onto a PVDF membrane (0.45-μm pore size, Procedure 1). The other set was separated by a 4–12% Bis-Tris NuPAGE gel, transferred by vertical transfer (25 V, 30 min) in solution onto nitrocellulose membrane (0.45-μm pore size, Procedure 2). Both membranes were subsequently treated in a similar manner using an anti-ACTH antibody to detect POMC-related proteins by enhanced chemiluminescence. Note the differential staining pattern for POMC, 23-KDa POMC intermediate. and ACTH between the two procedures for the same samples. B, basal, S, stimulated, L, lysate.

interest since subtle differences in the procedures can have significant effects on the results.

Alternate methods for the identification and quantification of processed hormones and neuropeptides include mass spectrometry (MS), which we already described and, when combined with the specificity of antibodies in the antibody capture procedure, can extend the power of this technique. Traditionally, radio-immunoassay (RIA) has been used to quantify specific peptides in a mixture. However, usually an antibody that identifies a mature processed neuropeptide also recognizes its precursor, although there are examples where RIAs can be exclusively specific for one species of the protein (e.g., mature human insulin, Linco Research, St. Charles, MO). While the antibody capture procedure for the protein chip reader utilizes the specificity of immune complexes followed by separation and identification by MS, reverse-phase HPLC followed by RIA first utilizes separation by liquid chromatography based on hydrophobicity and then identifies peptides by RIA in comparison to known standards. This procedure is useful when quantifying the relative amounts of precursor, intermediate peptides, and mature peptides in a given sample. In all likelihood, procedures for the analysis of peptides by HPLC have already been published; indeed, some companies have databases of procedures they can search upon request.

While static secretion experiments are common, the use of perifusion chambers provides a more dynamic measure of secretion events from cells. In brief, cells are placed in a chamber where they can adhere to or be retained in (as in the case of isolated pancreatic islets) and buffer is passed over them at a constant rate ($\sim$0.3 mL/min). The buffer is subsequently collected by a fraction collector, and the samples in each fraction can be assayed. The buffer can be easily switched by a valve to a stimulating buffer at a constant rate, or the stimulating buffer can be injected into the buffer system as a bolus. In both cases the cells' immediate response to the stimulating buffer can be measured. At the end of the experiment, the cells can be harvested as usual from the chamber. We obtained similar results in a static secretion experiment and a perifusion secretion experiment for the secretion of insulin-immunoreactive proteins from isolated pancreatic islets from wild-type (WT) and CPE KO animals, demonstrating the complimentarity of the two procedures.

Pulse-Chase Labeling Analysis

The pulse-chase paradigm allows analysis of secretion of newly synthesized proteins. The procedure for pulse-chase labeling was described earlier. The cell extract and medium collected during the chase period can be frozen until analyzed later or can be processed immediately for IP, as described in the next section. After IP, followed by gel electrophoresis, the dried gel containing the immunoprecipitated radioactive proteins is exposed to a phosphorous screen overnight, which is then recorded by a PhosphorImager. Typically, data from the PhosphorImager, representing the quantification of the radioactive bands on the gel in arbitrary units, can be analyzed by ImageQuant software (Molecular Dynamics).

In the analysis of the pulse-chase data, two important issues need to be considered: (1) For quantification, the amount of radioactivity in the prohormone or proneuropeptide and their processed products needs to be corrected for the number of radioactive amino acids present in the protein/peptide. For example, in the POMC molecule, there are five methionine residues (not including the leader sequence), but only one is in the processed product, ACTH. On an equimolar basis, therefore, the band intensity of POMC should be five times that of ACTH. Hence, only when the band intensities are corrected for the number of radioactive amino acids in the molecule can an accurate computation of the degree of processing of the precursors be made. Once the band intensity of the processed products has been corrected for the number of radioactive amino acids relative to the precursor, the amount of radioactivity (in arbitrary units) in the precursor and each of the processed product bands is expressed as a percentage of the total radioactivity (in arbitrary units) in the pro- and processed neuropeptides in cells plus medium. This allows for comparison between experiments where the degree of labeling differs due to differences in cell numbers.

(2) Comparison of the total radioactivity incorporated after the pulse ($T = 0$), versus the counts recorded after the chase period needs to be examined to establish possible degradation of the protein during the experiment. From a secretion assay using a pulse-chase paradigm, the $T = 0$ sample represents the total incorporation of radioactive amino acids into the protein of interest in the cell during the pulse period. For calculation of radioactivity at the end of the chase periods, the radioactivity in the chase medium and that in the cells harvested after the chase periods are summed. If the amount of total radioactivity recovered from the chase medium and remaining cells is less than the total counts incorporated into cells at $T = 0$, the difference would be interpreted as the amount of precursor/peptide degraded during the experiment. This type of analysis is particularly important when dealing with mutant proteins, as they are more likely to be degraded. Degradation needs to be factored into the interpretation of the data in determining the fate of the newly synthesized protein of interest (i.e., degradation, secretion, and storage).

Factors Influencing Secretion

Overexpression

Expression of transgenes in cells in culture has allowed the study and understanding of many cellular processes. The level of expression, however, is an important point to consider. Most of the mammalian expression vectors used today are driven by the cytomegalovirus (CMV) promoter, although other viral promoters are used in viral delivery methods, as we mentioned. The CMV promoter provides a high level of continuous expression of the transgene; however, consider the impact this has on the cell. Under steady-state conditions prior to expression, the cell is balanced in its cultured conditions. However, the onset of high expression of a protein provides the cell with a challenge: to recognize and use the protein as it normally would and then to deal with the rest that the cell does not or cannot use. In some cases the protein may accumulate in the cell and become toxic, while in other cases the excess protein can get degraded or be secreted extracellularly.

Overexpression becomes a problem in studying prohormone or proneuropeptide trafficking, as both the regulated and constitutive secretory pathways can become saturated. In the case of nerve growth factor (NGF), which is normally secreted via the constitutive pathway, overexpression in AtT20 cells resulted in its presence in the RSP *(15)*. On the other hand, proteins destined for the RSP, when highly expressed, can saturate this pathway, accumulate in the trans-Golgi network (TGN), and be secreted via the constitutive secretory pathway *(16)* (Fig. 5). For neuropeptides, the consequences of this are twofold. First, the increased constitutive secretion would result in the lack of processing of the proneuropeptides since they do not get into the granules of the RSP where processing occurs; second, the relative difference between constitutive versus regulated secretion would be reduced, indicating that the proneuropeptide was not sorted efficiently to the RSP, which may not be the case. All of these events (accumulation, degradation, saturation) as a result of overexpression may have an effect on the cell as a whole, as each event might require the up- or downregulation of components in those events, as is the case when the level of expression of chromogranin A (CgA) is modified in model (neuro)endocrine cell lines *(17)*. To minimize this, a balance should be made to express the smallest amount of protein that would allow sufficient detection by the experimental

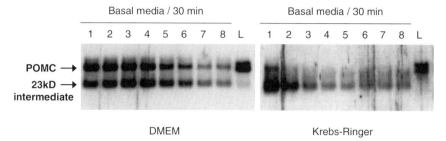

Fig. 5. The effect of buffers on the nonregulated secretion of POMC from AtT20 cells. AtT20 cells were incubated in either DMEM or Krebs–Ringer–HEPES (KRH) buffer for eight consecutive 30-min periods. After each period the medium/buffer was removed and prepared for Western blot analysis along with the cell extract from the cells (L) at the end of the experiment. POMC and its 23-KDa intermediate are secreted at a constant rate in DMEM, while in the KRH buffer POMC stopped being secreted within the first 30 min. The 23-KDa intermediate was also significantly reduced and then continued to be secreted at a low rate throughout the experiment.

assay used. A simple titration of the plasmid should be done to test this. Alternatively, inducible expression from plasmid or viral vectors using chemicals such as doxycyclin on the tetracycline promoter system has proven useful, although this requires more time to set up. These considerations would minimize the impact that overexpressed proteins might have on cellular processes, especially processes the experiment was designed to study.

State of the Cells

As with all cell culture systems, the longer the cells are cultured, the more difficult they are to work with. They begin to grow slower, detach more easily, and may begin to develop blebs on the cell body. It is important, therefore, to use cells that appear normal by examination at the light microscopic level in addition to performing a functional assay as an internal control. In the case of neurons and (neuro)endocrine cells, a functional assay with respect to neuropeptide trafficking would be the analysis of an endogenous protein as a marker for the regulated secretory pathway. CgA or Secretogranin II has been widely used *(6,17)*, although cell lines derived from endocrine tissues should be analyzed for their endogenous prohormones. Function can be assessed by performing a secretion assay and analyzing the relative levels being released under constitutive versus regulated conditions, or it can be observed by the correct immunofluorescent localization within the cell (see the upcoming section on the histochemical approach). Correct and consistent secretion behavior and cellular localization can be monitored throughout the experiments, and deviation from these expected results would be an indicator that the state of the cells might have changed.

When a cell line is obtained that can be demonstrated to have correct trafficking, secretion, and cellular localization of an endogenous marker, it should be expanded into 10 T-75-cm flasks and frozen for long-term storage in liquid nitrogen. One vial is then expanded into another 10 flasks, and 9 of these frozen, while the 10th one is used for experiments. These cells are used for 15–20 passages, after which a fresh vial is used from the remaining 9. When all 9 are used, a second vial from the original 10 flasks is used, expanded into another 10 from which 9 are frozen and the 10th one used. This should continue until the original cells are used and should ensure that the cells used at the beginning of a project are not too different from the cells used at the end.

Buffer System

Many secretion experiments have been carried out using a Krebs–Ringer-based buffer system *(18,19)*. We have found that this may not be appropriate for comparisons of relative secretory pathways between wild-type (WT) and mutant forms of prohormones and proneuropeptides. During a secretion experiment, which includes rinses, pre-equilibration, and the actual collection of basal or stimulated medium, the cells are being deprived of cofactors and amino acids. This is similar to the metabolic labeling process (described earlier), whereby the cell is starved of the amino acid that is to be used for the labeling, usually methionine (Met). Under these conditions, the absence of Met reduces protein synthesis, as the mRNA is halted within the ribosome until radioactive Met is introduced. Since the Krebs–Ringer buffer is devoid of all amino acids, protein synthesis becomes and remains minimal. The consequence of this is described in the example that follows, comparing the unregulated secretion behavior of POMC from AtT20 cells using either DMEM or Krebs–Ringer buffer.

AtT20 cells, a mouse corticotroph cell line that expresses POMC, were grown to 80–90% confluency in 6-cm dishes with DMEM supplemented with 10% FBS and 1X penicillin/streptomycin. The cells were then rinsed twice with either DMEM or Krebs–Ringer–HEPES buffer (KRH)(129 mM NaCl, 5 mM NaHCO$_3$, 4.8 mM KCl, 2.8 mM glucose,1.2 mM KH$_2$PO$_4$, 1.2 mM MgCl$_2$, and 1 mM CaCl$_2$, 10 mM HEPES, pH 7.4). The cells were then incubated at 37 °C with 1 mL of either DMEM or KRH buffer for eight consecutive 30-min periods. After each 30-min period, the buffer was collected and centrifuged for 5 min at 1,000 x g in a microcentrifuge, and 800 µL of the supernatant were added to fresh tubes containing 8 µL of 10% SDS at room temperature. The samples were mixed with the SDS and immediately frozen on dry ice and saved at –20 °C until analyzed. The cells were rinsed with ice cold PBS twice and harvested with 1 mL cell lysis buffer (M-per mammalian protein extraction buffer, Pierce). After centrifugation at 15,000 x g for 20 min, the supernatant was transferred to a new tube and frozen.

Analysis of the POMC material in the samples was performed by Western blot using rabbit anti-ACTH antiserum (DP4, 1:2000 dilution), generated in our laboratory, which cross-reacts with POMC, the 23-kDa intermediate and the glycosylated and non-glycosylated forms of ACTH. The samples were thawed on ice and mixed, and 20 µL of each were prepared for SDS-PAGE and Western blot. POMC and its 23-kDa intermediate were secreted from AtT20 cells that were incubated with DMEM. The amount detected in the medium remained elevated and constant for over 2 h, and the ratio between the two forms remained similar (Fig. 5, left panel). In contrast, the experiment carried out with the Krebs–Ringer buffer yielded a different secretion pattern of POMC and its 23-kDa intermediate. After the first 30-min incubation, the amount of POMC was significantly reduced and remained so throughout the experiment. The 23-kDa intermediate was present in the first 30 min but diminished significantly after the first hour to a low basal level for the remainder of the experiment (Fig. 5, right panel). The POMC doublet remaining in both sets of cells after the 4 h represents the cellular forms of the precursor.

These results demonstrate that the behavior of constitutive secretion of POMC from AtT20 cells depends upon the buffers used. In the absence of protein synthesis, in an amino acid-deficient buffer, the POMC that is in transit in the cell prior to the experiment continues until it has reached its final destination. In this case it is secreted to a significant extent within 30 min of being deprived of amino acids, via the constitutive or constitutive-like secretory pathways. Since rinses and pre-equilibration incubations are commonly performed and discarded prior to a secretion experiment, the subsequent basal secretion sample would not represent the true basal secretion, since the cells would not be under steady state.

Immunochemical Approach

A complementary procedure to that of secretion assays in the study of cellular trafficking and processing is that of immunocytochemistry (ICC). The following sections describe basic procedures and provide information such as the use of intracellular markers and possible artifacts for developing an appropriate protocol for the analysis of the cellular localization of a protein of interest.

Preparation of Cells for Immunolabeling Studies

Since (neuro)endocrine cells develop processes or neurites where granules/synaptic vesicles of the RSP are stored, an appropriate amount of time is needed for newly plated cells to grow in order to establish normal morphology and neurite outgrowth before ICC is carried out. Also important is the density of the cells when the ICC is performed. One should aim for a confluency of not more than 60–70%. Typically allow 24–48 h for normal morphology to develop and to achieve the required confluency, although the time needed may depend on the specific cell type. Cells may be plated on sterile microscope cover glasses (Fisher Scientific, PA, USA) or on special chambered slides. Advantages for using cover glasses are (1) an equal and uniform distribution of the cells can be obtained for each experimental set, since many cover glasses onto which the cells are plated can be placed into one cell culture dish, (2) a small amount of diluted primary and secondary antibodies (30–50 µL) is enough to cover the cells on one cover glass (12 mm in diameter), and (3) cover glasses are less expensive than the chambered slides. Pretreatment of the chambered slides or cover glasses with 0.01% poly-L-lysine (AtT20 and Neuro2A cells) or 0.25–0.5 mg/mL collagen (for PC12 cells and primary neurons) will increase the attachment of cells to the surface and prevent their loss during subsequent washing steps in the ICC procedure. To do this, place several sterile cover glasses into a culture dish and add enough solution to cover them. Incubate for 1–2 h at 37 °C, then remove the solution, and rinse the dish a few times with regular growth medium (for example: DMEM, 10% FBS) or sterile water in order to remove any excess poly-L-lysine or collagen. The culture dish is now ready for plating cells.

Immunolabeling Protocol

The cover glasses with cells attached are transferred into a clean culture dish, rinsed twice with PBS, and fixed with either 4% paraformaldehyde (PFA) in PBS for 10 min at room temperature (RT) or with –20 °C cold methanol for 5 min in a –20 °C freezer. The cells are then rinsed four times for 5 min each with PBS and permeabilized with 0.1% Triton-X-100 in PBS for 10 min at RT. Fixation with ice cold methanol does not require this permeabilization step. The cells are then rinsed again as before, followed by incubation for 30 min at RT in blocking solution (1% BSA in PBS). As a blocking solution one can also use 3–5% normal horse or goat serum. For primary antibody incubation, the cover glasses are placed (cells up) on parafilm inside a small tray using forceps with fine tips. A sufficient volume (30–50 µL) of primary antibodies, diluted in blocking solution, is then placed on top of the cover glass to cover the cells. To maintain humidity during the subsequent 1-h incubation at RT, a moist strip of cut sponge or several layers of moist filter paper are placed at one end of the tray, which is then covered. After 1 h, transfer the cover glasses from the parafilm into separate culture dishes (to avoid cross-contamination of the primary antibody) and wash three times for 5 min in wash buffer (PBS containing 0.1% BSA and 0.1% Tween-20) at RT. Then transfer the cover glasses to clean parafilm in the tray and incubate with the secondary antibodies diluted in blocking solution for 1 h at RT. After washing three times for 5 min each in wash buffer and rinsing once with water; apply 10–20 µL of mounting medium to a clean microscopic slide and place the cover glass with cells upside down on top of the mounting medium while avoiding air bubbles. Keep the slides overnight in a dry and dark place for immediate analysis. For longer storage, keep the slides refrigerated in the dark.

Choice of Antibodies for Multiple Labeling

The strength of immunofluorescent localization is greatly enhanced when the procedure is performed in combination with two or three primary antibodies simultaneously to allow the identification of the compartment where the protein under study is localized by comparison with a well-characterized marker protein for a specific organelle. However, this type of double and triple labeling is only possible if the primary antibodies were raised in different host species, i.e., rabbit, mouse, and chicken, or rabbit, mouse, and goat. Optimum antibody concentrations are experimentally established, first by starting with the manufacturer's specifications. All primary antibodies are combined in one tube at their respective dilutions and incubated with the cells. For example, for labeling endoplasmic reticulum (ER), the Golgi apparatus, and dense core granules in AtT20 cells, a mouse pituitary corticotroph cell line, mix the primary antibodies: chicken anti-Calreticulin, an ER resident protein (Chemicon International, Temecula, CA, USA), at 1:500 dilution, rabbit anti-GRASP65, a Golgi marker (Proteus Biosciences, Ramona, CA, USA), at 1:1000 dilution and mouse anti-ACTH, a dense core granule marker (Abcam, Cambridge, MA, USA), at 1:1000 dilution.

In order to choose the correct fluorescently labeled secondary antibodies, one has to consider two things: First, the secondary antibodies are species-specific; and second, appropriate fluorophores are used, which can be observed separately. The best choice is to use secondary antibodies from the same host (goat anti-mouse, goat anti-rabbit, and goat anti-chicken). If it is not possible, make sure that there is no cross-reactivity between secondary antibodies. For example, mixing goat anti-rabbit and rabbit anti-mouse secondary antibodies will result in binding between them and labeling one secondary antibody with another secondary antibody, the result of which gives an incorrect interpretation of 100% co-localization between the two proteins. It is also important to ensure that all secondary antibodies are pre-absorbed against other hosts (i.e., goat anti-mouse antibody must not recognize rabbit IgGs). This information is usually provided by the manufacturer and should be mentioned on the specification sheet. We have had the best experience with secondary antibodies conjugated with Alexa dyes from Molecular Probes, Inc. (Eugene, OR, USA).

Fluorophores attached to secondary antibodies have defined spectroscopic parameters: excitation and emission spectra. In order to distinguish secondary antibodies on the microscope, emission spectra of their fluorophores should not overlap. A popular choice of fluorophores for double labeling includes Alexa-488 (green; fluorescence emission 519 nm) and Alexa-568 (red; fluorescence emission 603 nm). For triple-labeling experiments, one may consider including Alexa-350 (blue; fluorescence emission 442 nm) or Alexa-647 (far red; fluorescence emission 668 nm) as well. The appropriate set of excitation/emission filters on the microscope must be installed. Remember that the excitation/emission spectrum of green fluorescent protein (GFP) almost coincides with Alexa-488; the spectra of dsRed coincide with Alexa-568 and the spectra of DAPI coincide with Alexa-350. Therefore, the choice of label on the secondary antibody is critical; for example, when co-localizing transfected GFP-fusion proteins with endogenous markers, avoid using Alexa-488.

Quantification: Confocal Microscopy and Subcellular Fractionation

While direct observation of ICC slides reveals a qualitative picture as to the general cellular localization of prohormones/proneuropeptides and their processed products, ICC can also provide quantitative data. With respect to the sorting efficiency of an exogenously expressed protein to the RSP in (neuro)endocrine cells, this is a useful mode of analysis. This is done by comparing the immunofluorescent staining pattern of the protein of interest with the pattern of an endogenous marker used to identify dense core secretory granules of the RSP. Using single-section images captured by confocal microscopy, we assess a minimum of 100 transfected cells that have long processes (where mature secretory granules are mostly localized) and count how many of these cells have co-localization between the transfected protein and the mature granule marker along and at the tips of the processes. Using numbers from at least three independent experiments, a quantitative and representative set of data can be obtained demonstrating the percentage of cells with co-localization of the transfected protein in the granules of the RSP. A reduction in this percentage would be interpreted as a decrease in the efficiency of trafficking of the protein of interest to the RSP.

Quantification of ICC slides, however, can be subject to observer bias. Therefore, to support the observations of the ICC, we undertake a biochemical approach using sucrose equilibrium density gradients. The assumption being made when comparing a wild-type and mutant prohormone is that if the mutant is not sorted efficiently to the mature granules of the RSP, its presence in the gradient fractions corresponding to the mature granules would be quantitatively reduced. Conversely, its presence in a Golgi fraction is predicted to be increased, although this need not be the case exclusively. As an example that follows, we have demonstrated a differential localization of WT human growth hormone (hGH) and a mutant hGH, that has recently been described (20), between Golgi-containing and mature granule-containing fractions from sucrose gradients. AtT20 cells were grown to 80–85% confluency on 150-mm dishes and transfected with WT hGH or mutated growth hormone (mutGH) pcDNA3.1 constructs using Lipofectamine2000 (Invitrogen) according to the manufacturer's protocol. After 24 h, the cells were harvested in 1 mL of homogenization buffer [0.32 M sucrose, 150 mM NaCl, 1 mM EGTA, 0.1 mM $MgCl_2$, 20 mM HEPES, 1X inhibitor cocktail (Roche), pH 7.2]. The cells were broken by passage through a 27-gauge needle for 10 strokes. The postnuclear supernatant (PNS) was collected after centrifugation for 5 min at 1,000 x g and layered onto a linear 0.6–1.6 M sucrose gradient that was made using a Gradient Master (Biocomp). The gradients were centrifuged in an SW28.1 rotor (Beckman) for 16 h at RCF_{avg} 92,500 x g (RCF_{max} 130,000 x g) at 4 °C. Following centrifugation, 16 × 1 mL-fractions were collected from the top using a fraction collector and analyzed by Western blot after fractionation through a 4–20% Tris/Glycine SDS-PAGE gel. The PVDF membrane was probed with anti-growth hormone, anti-Vti1a (a Golgi marker), and anti-ACTH (a mature granule marker) antibodies. Visualization of the proteins was by indirect detection of secondary antibodies labeled with fluorophores that emit in the infrared region of the spectrum using the Odyssey Infrared Imaging System (Licor Biosciences). Individual bands were quantified using the Odyssey software and imported into Microsoft Excel for processing. The data clearly show that, compared to WT hGH, the mutant is significantly reduced in the ACTH-containing fractions, while it is increased in the Vti1a-containing fractions (Fig. 6). Thus, the differential localization of WT and mutant hGH observed by ICC (20) is confirmed by the subcellular fractionation studies.

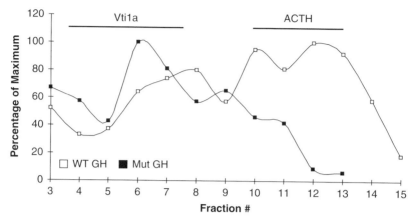

Fig. 6. Analysis of the cellular distribution of WT and mutant human growth hormone (hGH) by sucrose equilibrium centrifugation. Mutant hGH was previously observed by immunocytochemistry to be localized to granules of the regulated secretory pathway (RSP) in AtT20 cells to a lesser extent than WT hGH *(20)*. To support this observation, cells expressing both proteins were subjected to sucrose equilibrium centrifugation on a gradient designed to separate the Golgi apparatus and mature granules of the RSP. Fractions from the gradient were analyzed by Western blot for ACTH (granule marker) and Vti1a (Golgi marker) and hGH. In support of the ICC, mutant hGH was distributed preferentially in a Golgi fraction, and to a lesser extent in the granules, compared to WT hGH, which was localized preferentially in the granules, as well as in the Golgi, but to a lesser extent.

Concluding Remarks

In this chapter, we have provided the reader with a strategy and state-of-the-art techniques used for studying the processing, trafficking, and secretion of neuropeptides. Details of protocols can be obtained from *Current Protocols in Cell Biology* or from the papers cited here. More importantly, we have tried to share our own experiences to illustrate how differences in data can be obtained from using different procedures that one would not have expected to have had any effect. For example, the choice of secretion medium in release studies made a significant difference in the products released from the cell. Also, we have attempted to highlight caveats in data interpretation depending on the experimental design, technical strategies to increase sensitivity, and the need to use subcellular fractionation techniques to confirm immunohistochemical localization of neuropeptides. We hope the reader will find this chapter helpful in their experimental design to study proneuropeptide processing, trafficking, and secretion.

References

1. Dhanvantari, S., Shen, F.-S., Adams, T., Snell, C.R., Zhang, C., Mackin, R.B., Morris, S.J., and Loh, Y.P. Disruption of a receptor-mediated mechanism for intracellular sorting of proinsulin in familial hyperproinsulinemia. *Mol. Endocrinol.*, 2003; **17**(9): 1856–1867.
2. Yanik, T., Dominguez, G., Kuhar, A.J., Miraglia Del Guidice, E., and Loh, Y.P. The Leu34Phe ProCART mutation causes CART deficiency leading to obesity in humans. *Endocrinology*, 2006; **147**(1): 39–43.
3. Egan, M.F., Kojima, M., Callicott, J.H., Goldberg, T.E., Kolachana, B.S., Bertolino, A., Zaitsev, E., Gold, A., Goldman, D., Dean, M., Lu, B., and Weinberger, D.R. The BDNF val66met polymorphism affects activity-dependent secretion of BDNF and human memory and hippocampal function. *Cell*, 2003; **112**(2): 257–269.
4. Yamanaka, R., and Xanthopoulos, K.G. Development of improved Sindbis virus-based DNA expression vector. *DNA Cell. Biol.*, 2004; **23**(2): 75–80.
5. Kohno, A., Emi, N., Kasai, M., Tanimoto, M., and Saito, H. Semliki Forest virus-based DNA expression vector: Transient protein production followed by cell death. *Gene Ther.*, 1998; **5**(3): 415–418.
6. Lou, H., Kim, S.-K., Zaitsev, E., Snell, C.R., Lu, B., and Loh, Y.P. Sorting and activity-dependent secretion of BDNF require interaction of a specific motif with the sorting receptor carboxypeptidase E. *Neuron*, 2005; **45**(2): 245–255.
7. Caputo, E., Moharram, R., and Martin, B.M. Methods for on-chip protein analysis. *Anal. Biochem.*, 2003; **321**(1): 116–124.
8. Diamond, D.L., Zhang, Y., Gaiger, A., Smithgall, M., Vedvick, T.S., and Carter, D. Use of ProteinChip array surface enhanced laser desorption/ionization time-of-flight mass spectrometry (SELDI-TOF MS) to identify thymosin beta-4, a differentially secreted protein from lymphoblastoid cell lines. *J Am. Soc. Mass Spectrom.*, 2003; **14**(7): 760–765.
9. Petricoin, E.F., and Liotta, L.A. SELDI-TOF-based serum proteomic pattern diagnostics for early detection of cancer. *Curr. Opin. Biotechnol.*, 2004; **15**(1): 24–30.
10. Cawley, N.X., Zhou, J., Hill, J.M., Abebe, D., Romboz, S., Yanik, T., Rodriguiz, R.M., Wetsel, W.C., and Loh, Y.P. The carboxypeptidase E knockout mouse exhibits endocrinological and behavioral deficits. *Endocrinology*, 2004; **145**(12): 5807–5819.
11. Cool, D.R., and Hardiman, A. C-terminal sequencing of peptide hormones using carboxypeptidase Y and SELDI-TOF mass spectrometry. *Biotechniques*, 2004; **36**(1): 32–34.
12. Patterson, D.H., Tarr, G.E., Regnier, F.E., and Martin, S.A. C-terminal ladder sequencing via matrix-assisted laser desorption mass spectrometry coupled with carboxypeptidase Y time-dependent and concentration-dependent digestions. *Anal. Chem.*, 1995; **67**(21): 3971–3978.
13. Kristensen, P., Judge, M.E., Thim, L., Ribel, U., Christjansen, K.N.,. Wulff, B.S., Clausen, J.T., Jensen, P.B., Madsen, O.D., Vrang, N., Larsen, P.J.,

and Hastrup, S. Hypothalamic CART is a new anorectic peptide regulated by leptin. *Nature*, 1998; **393**(6680): 72–76.

14. Suzuyama, K., Shiraishi, T., Oishi, T., Ueda, S., Okamoto, H., Furuta, M., Mineta, T., and Tabuchi, K. Combined proteomic approach with SELDI-TOF-MS and peptide mass fingerprinting identified the rapid increase of monomeric transthyretin in rat cerebrospinal fluid after transient focal cerebral ischemia. *Brain Res. Mol. Brain Res.*, 2004; **129**(1–2): 44–53.

15. Mowla, S.J., Pareek, S., Farhadi, H.F., Petrecca, K., Fawcett, J.P., Seidah, N.G., Morris, S.J., Sossin, W.S., and Murphy, R.A. Differential sorting of nerve growth factor and brain-derived neurotrophic factor in hippocampal neurons. *J. Neurosci.*, 1999; **19**(6): 2069–2080.

16. Eaton, B.A., Haugwitz, M., Lau, D., and Moore, H.P. Biogenesis of regulated exocytotic carriers in neuroendocrine cells. *J. Neurosci.*, 2000; **20**(19): 7334–7344.

17. Kim, T., Tao-Cheng, J.H., Eiden, L.E., and Loh, Y.P. Chromogranin A, an "on/off" switch controlling dense-core secretory granule biogenesis. *Cell*, 2001; **106**(4): 499–509.

18. Irminger, J.C., Verchere, C.B., Meyer, K., and Halban, P.A. Proinsulin targeting to the regulated pathway is not impaired in carboxypeptidase E-deficient Cpefat/Cpefat mice. *J. Biol. Chem.*, 1997; **272**(44): 27532–27534.

19. Cowley, D.J., Moore, Y.R., Darling, D.S., Joyce, P.B., and Gorr, S.U. N- and C-terminal domains direct cell type-specific sorting of chromogranin A to secretory granules. *J. Biol. Chem.*, 2000; **275**(11): 7743–7748.

20. Wang, J., Cawley, N.X., Voutetakis, A., Rodriguez, Y.M., Goldsmith, C.M., Nieman, L.K., Hoque, A.T., Frank, S.J., Snell, C.R., Loh, Y.P., and Baum, B.J. Partial redirection of transgenic human growth hormone secretion from rat salivary glands. *Hum. Gene Ther.*, 2005; **16**(5): 571–583.

8

In vivo Techniques Quantifying Blood–Brain Barrier Permeability to Small Proteins in Mice

Weihong Pan, and Abba J. Kastin

Key Words: blood–brain barrier; peptides; cytokines; influx; efflux; transport; multiple-time regression analysis

Abstract

The blood–brain barrier (BBB) plays a crucial regulatory role in central nervous system (CNS) function and in communication between the CNS and the periphery. In addition to lipophilic molecules, many small proteins are now known to cross the BBB. Such recognition was expedited by techniques that quantify the influx and efflux of peptides and polypeptides across the BBB of the living animal. These methods are described in this review. Their use has enhanced our knowledge of the crucial link between the CNS and the rest of the body, and this has physiological and pathological implications as well as therapeutic potential.

Introduction

The blood–brain barrier (BBB), while restricting the entry of many toxins and other substances into the brain, provides a large neurovascular interface with regulatory functions for the optimal performance of the central nervous system (CNS). The main component of the BBB is a specialized microvascular system. Elsewhere in the body, the small blood vessels are lined by endothelial cells that are characterized by gaps between them that facilitate the rapid passage of nutrients. By contrast, microvascular endothelial cells in the CNS are closely connected with each other by tight junctions and are lined by a continuous basement

From: *Neuromethods, Vol. 39: Neuropeptide Techniques*
Edited by: I. Gozes © Humana Press Inc., Totowa, NJ

membrane lacking fenestration. This nonleaky feature is further reinforced by pericytes and the endfeet of astrocytes.

Techniques to determine rapid extraction of ions, water, and other smaller molecules from arterial blood were developed in the 1970s *(1–4)*. However, methods to evaluate the passage of slow-penetrating peptides and proteins across the BBB *in vivo* required further development. Now these sensitive methods can quantify the BBB passage of peptides (fewer than 100 amino acids) and polypeptides (100–200 amino acids). By these arbitrary definitions of small proteins, most cytokines, chemokines, and neurotrophins are considered polypeptides. For ease of communication in this review, the terms "peptides," "polypeptides," and "proteins" will be used interchangeably. *In vitro* techniques with cultured cells, although crucial to determining the mechanisms of intracellular trafficking, will not be covered in this review.

A major limitation of *in vivo* methods lies in potential interactions of the study substance with the peripheral circulation before it reaches the BBB. This includes protein binding and enzymatic degradation after intravenous (iv) injection. At the BBB, there is also enzymatic activity that can show regional differences *(5)*. These possible difficulties are somewhat obviated by *in situ* brain perfusion in a blood-free buffer.

After an iv bolus injection, the blood concentration of the exogenous protein decreases rapidly because of tissue distribution, metabolism, and renal excretion. This disappearance is corrected by mathematical modeling by multiple-time regression analysis, modified from the methods of Fenstermacher, Blasberg, and Patlak *(6–8)*. Such modeling creates a theoretical steady-state concentration for the circulation time of the test substance.

Multiple-Time Regression Analysis

Radioactive Labeling (Radiolabeling)

Why Radiolabel?

Detection by radioimmunoassay (RIA), enzyme-linked immunosorbent assay (ELISA), chromatography (e.g., Sephadex, high-performance liquid chromatography: HPLC, sodium dodecylsulfate-polyacrylamide gel electrophoresis: SDS-PAGE), or even mass spectroscopy encounters the difficulty of distinguishing the injected protein of interest from endogenous production of the same substance. Except in the situation with delta sleep-inducing peptide (DSIP), for which an unusual antibody was generated that recognized almost all (8/9) of the sequential amino acids *(9)*, the additional danger for RIA and ELISA is the probability of merely measuring a metabolic fragment since most antibodies recognize only three or four amino acids, and even these may not be adjacent. Of course, if the antibody used for RIA or ELISA is species-specific, such as we used for detection of human insulin in the mouse *(10)*, then useful information can be obtained about BBB penetration without use of a radiolabeled peptide/polypeptide. Although fluorescent labeling has been increasingly popular in the past decade, the necessity of tissue homogenization, together with a relatively lower specific activity, makes the technique cumbersome and less sensitive in kinetic assays of whole brain tissue.

Radioactive Iodine

The most common method involves a radiolabeled protein. Use of radiolabeled substances greatly reduces the amount required for detection. The method is old, standard, and still useful *(11)*. It is much simpler, faster, and more sensitive to label the protein being tested with a gamma-emitting source—usually ^{125}sodium iodide (I)—which results in high specific activity. ^{131}I can be used, but it has a shorter half-life (8 days vs. 60 days) and more penetrating radiation. Proteins with a tyrosine (Tyr) are readily labeled with ^{125}I; a histidine (His) can also be tagged in this way, particularly at a higher pH and if sterically unhindered, but this usually is more difficult. Most investigators use chloramine-T (*n*-chloro-para-toluene sulfonamide sodium salt) to release radioactive elemental iodine. Being a strong oxidizing agent, however, chloramine-T can damage the protein. Care must be taken to ensure that only the undamaged peptide is used for the studies. This usually is achieved by chromatographic separation. If the supplier of the peptide, which should be carrier-free, provides an HPLC elution profile, then the use of that gradient accompanied by UV readings at 280 and 214 can identify the correct undamaged fraction.

Milder forms of labeling with radioactive iodine entail Iodogen beads or the Bolton–Hunter reagent. The Iodogen method, involving immobilized lactoperoxidase *(12)*, has the advantage of avoiding direct contact between peptide and oxidizing agent, but the beads are made of polystyrene, which may adsorb the peptide, and the yield tends to be low. The reaction is stopped by removal of the labeled compound rather than by addition of the reducing agent sodium metabisulfite that is commonly used for the chloramine-T method.

The Bolton–Hunter method involves an active intermediate that is iodinated and then conjugated under mild conditions with the peptide, also avoiding direct contact of the reagent with the peptide *(13)*. The reagent, the iodinated N-hydroxy succinimide ester of p-hydroxyphenyl propionic acid, is prepared with a modified chloramine-T procedure. Being a non-oxidative procedure, it can be useful for polypeptides containing Met or Cys that need to be kept in a reduced state. The Bolton–Hunter procedure attaches residues similar in structure to Tyr to amino groups of the test peptide, but it is not widely used. In at least one situation, the Bolton–Hunter method, unlike the chloramine-T method, resulted in almost total loss of activity *(14)*. For small proteins lacking a Tyr or His, a frequently used technique involves synthesis with an added Tyr, usually at the N-terminus. In many situations this additional amino acid, like iodination, does not seem to affect biological activity.

Precautions

It should be remembered that peptides containing Trp, Met, or Cys are easily oxidized, resulting in altered activity. Also, many peptides, especially basic ones, stick to glass (negatively charged) and some synthetic surfaces. This is usually prevented by the addition of a small amount of albumin, detergent, or acidification (except for sulfated peptides, which do better at a basic pH). It is preferable to use polypropylene labware rather than polystyrene. We found that coating glass surfaces with silicone can increase nonspecific adherence in some situations.

Tritium

A more costly way to radiolabel a peptide that lacks a Tyr (or His) is to label it with tritium (or ^{14}C), a low-energy, beta-emitting radionuclide with a half-life of 12.5 years. The result, especially with carbon labeling, frequently has a lower specific activity than with iodination. Nonspecific generalized tritiation, although relatively less expensive, is not as desirable as selective tritiation, which usually involves synthesis of a precursor. Generalized exchange tritiation tends to cause radiation-induced decomposition and results in random labeling with very low specific activity and many impurities.

Derivatized tritiation is the method of choice for adding a tritium, but even this difficult and expensive technique can result in substantial changes in the conformation of the peptide and requires extensive purification. Moreover, measurement of radioactivity is more tedious for tritiated than iodinated proteins, requiring the tissues to be digested, quenched, and finally counted in a liquid scintillation counter. This can involve overnight incubation in a shaking water bath at 50 °C, addition of hydrogen peroxide to the digested tissue and a second overnight incubation (at 4 °C) for decoloration, and finally a third incubation with scintillation cocktail in the dark for 24 h to convert beta energy into light photons. Since the amount of quenching can vary from sample to sample, it may be necessary to estimate the efficiency of counting for each individual sample. Even though iodine is much larger than tritium, we found no significant differences in BBB transport with two different peptides each labeled with iodine or tritium *(15,16)*.

Experimental Design

Controls

To ensure that the influx of the test substance is not caused by disruption of the BBB, a control for the vascular space is often included in the study. This is particularly necessary when it is not certain whether the test substance has vasoactive effects. Bovine serum albumin (BSA, MW 66,000) is usually used and, if radiolabeled with ^{99m}Tc or ^{131}I, can be injected together with the ^{125}I-labeled protein of interest so that both can be counted at the same time in a dual-channel gamma counter. The half-life for the gamma emission of ^{99m}Tc, however, is only 6 h. Other suitable controls include the denatured protein of interest (e.g., incubated at 80 °C in a water bath for 10 min followed by immersion in ice for 5 min), a mixture of the component amino acids, lactoalbumin (MW 14,200), or smaller markers such as inulin (MW 5,200). If the experiment uses gamma radiation for detection, then inulin, which cannot be iodinated, usually is injected into separate mice. Alternatively, if the sample contains a beta-emitting compound (either protein or control substance) together with a gamma-emitting compound, measurement in the gamma counter could be made first; then, after a wait for complete gamma decay, the other compound can be measured in a liquid scintillation counter without concern about crossover.

Initial Test of Stability

The first step should be determination of the stability of the protein in question. In general, peptides are rapidly metabolized in blood. It would be difficult to interpret the experimental data if most of the peptide did not remain intact in the circulation for 10 min. HPLC of the labeled peptide/polypeptide (or SDS-PAGE of the polypeptide) entering the CNS and the proportion remaining intact in both blood and brain, therefore, are necessary to determine whether there is permeation of the intact protein across the BBB.

Saturability

The two main types of entry into the CNS are passive diffusion and saturable transport. Passive diffusion is dependent on the physicochemical properties of the protein, including lipid solubility in the plasma membrane, hydrogen bonding, conformation, and, to a relatively minor extent, molecular size *(17–19)*.There are many examples of larger molecules entering the brain faster than smaller molecules *(20)*. Saturable transport represents regulated permeation of the BBB that can be modulated under physiological as well as pathological conditions. The most physiological method of testing for saturation is to include an excess amount of unlabeled peptide in the same solution injected intravenously with the radiolabeled peptide. Obviously, another group in this design must be injected only with the radiolabeled peptide.

Tissue Sample

Regional differences in BBB permeability have been found in normal brain and spinal cord, and regional differences in the regulation of permeability can reflect the susceptibility of specific CNS areas to pathological insults *(21–23)*. Thus, in addition to evaluation of total brain and spinal cord uptake of tracers, it is possible to carefully dissect these tissues to examine regional differences. This is particularly useful when there is focal pathology. For instance, the rodent model of ischemic stroke—transient middle cerebral artery occlusion (tMCAO)—produces a distinct ischemic core as well as penumbrae at the border of the infarct area. These regions show different degrees of impairment and regulation of BBB permeability, and the changes in permeability and the transport systems in turn affect the process of neuroregeneration *(24)*. Another example of regional sampling involves the circumventricular organs (CVOs). Although it is a misconception to consider that CVOs represent ready access to the entire brain *(25,26)*, the use of cortical areas, which intrinsically are devoid of CVOs, avoids this potential problem. Studies with interleukin-1α have quantified the minimal amounts entering the brain through the CVOs *(27,28)*.

Procedure

The test substance (e.g., radiolabeled peptide) and reference compound (e.g., albumin) are injected together as a bolus into the jugular vein of anesthetized adult mice in a total volume of 100 µL/mouse, diluted in lactated Ringer's solution containing 1% BSA. Any hemodynamic changes caused by this small volume usually do not affect the permeability measurement during the short time of the study. At designated time points after injection, arterial blood is collected from the right carotid artery immediately before decapitation.

Calculations

K_i and V_i

The brain-to-serum ratio of the radioactivity for each isotope is expressed in unit weight of tissue and unit volume of serum (cpm or dpm/g)/(cpm or dpm/mL), or mL/g. The serum radioactivity (cpm or dpm in 50 µL of serum) at each time point is plotted in a semi-logarithm curve against time. The exposure time (Expt) at time t is calculated as the integral of serum radioactivity divided by the radioactivity at time t, thus generating the theoretical time for the steady-state concentration of circulating labeled protein after correction for disappearance from blood. It may seem paradoxical that a brain-to-blood ratio is expressed as mL/g rather than g/mL, but the reason for this becomes apparent by consideration of the calculation. The (cpm/g brain)/(cpm/mL blood) can be rearranged to (cpm)(mL blood)/(cpm)(g brain) so that the cpm in the numerator and denominator cancel.

The unidirectional influx rate K_i (mL/g-min) and the initial volume of distribution V_i (mL/g) are derived from the following linear regression: Brain-to-serum ratio of radioactivity = K_i.(Expt) + V_i. Thus, the combined formula is

$$A_m/C_{pt} = K_i \cdot \int_0^t C_p(t)d\tau/C_{pt} + V_i,$$

with A_m being the amount of radioactivity in the brain at time t, C_p the concentration of arterial serum, C_{pt} the amount of radioactivity in serum at time t, and τ the dummy variable for time.

Linear Regression

GraphPad Prism statistical software estimates the linearity of the regression correlation. The slope of the regression line between tissue-to-serum ratios and exposure time is K_i, the unidirectional influx constant reflecting entry rate. Experimental time should not be confused with exposure time, which corrects for decay of radioactivity over time. The K_i calculated from exposure time is lower than that derived from the linear regression between the tissue-to-serum ratio and experimental time. As mentioned previously, only the linear portion of the curve is used. The V_i, essentially the intercept of the linear regression line with the ordinate, reflects the initial volume of distribution in the brain or other tissues being measured.

Comparison with Permeability Coefficient

The measure of BBB influx described in this review, the K_i, is equivalent to the permeability coefficient × surface area product (PS value) that also has been used. Flow rate (F/unit mass of tissue) is stable and permeability low for most small non-vasoactive proteins. Thus, with the extraction fraction (K_i/F) being <0.1, the equation PS = -F ln[1 - (K_i/F)] essentially becomes PS • K_i.

In situ Perfusion

One possible explanation for the lack of significant influx of a peptide could be because it is bound to serum proteins.

Perfusion in a blood-free buffer obviates this problem. As with iv injection, the radiolabeled protein to be tested is administered together with radiolabeled albumin. Albumin is the reference tracer to ensure that there is no disruption of the BBB caused by the procedure. Matters to be taken into consideration include perfusion buffer, delivery method, perfusion speed, and duration. For mouse studies, we usually deliver a physiological buffer resembling artificial cerebrospinal fluid through the left cardiac ventricle. To reduce the amount of tracer needed, the abdominal aorta is clamped so that perfusion only occurs in the upper part of the body. To create an outflow, either the right cardiac atrium or the jugular veins are severed. Oxygenated perfusate is delivered at a rate of 2 mL/min at various times up to 10 min, although 5 min is usually used. Each mouse receives a prewash for 2 min to clear the vasculature and a postwash of 1 min to remove any radiolabeled material that has not entered the tissue. At each experimental time, the mouse is decapitated and the brain-to-perfusate ratio of radioactivity per g of brain is determined. Tissue influx is calculated from the slope of the regression line for the brain-to-perfusate ratio of radioactivity vs. time. The composition (g/L) of the buffer, adjusted to pH 7.4, is NaCl: 7.19, KCl: 0.3, $CaCl_2$: 0.28, $NaHCO_3$: 2.1, HK_2PO_4: 0.16, $MgCl_2/6H_2O$: 0.37, D-glucose: 0.99, 1% BSA.

Compartmental Distribution

Background

Simple measurement of radioactivity in the brain includes that located in the vasculature which does not reflect permeation into parenchyma. No cellular element of the brain is more than 50–70 μm from a capillary. The capillary depletion method (29), as modified for mice (30), separates the cerebral microvasculature from tissue parenchyma. A high K_i may not necessarily reflect actual entry into brain parenchyma, such as occurs with the substantial capillary association of transforming growth factor α (TGFα) (31). Real influx requires demonstration of the presence of the injected protein in the parenchymal compartment of the brain.

Gamma Glutamyl Transpeptidase

Gamma glutamyl transpeptidase (γGT) is abundantly present on the apical surface of endothelial cells. It serves as a reliable marker for the microvasculature of the BBB and therefore effective separation of the capillaries from the parenchyma. We compared the yield and efficiency of samples obtained with a Beckman TL-100 ultracentrifuge and the smaller, less expensive Beckman microfuge 22R centrifuge by analyzing γGT activity and protein concentrations in both pellet and supernatant *(32)*. Both centrifuges use swinging-bucket rotors. The higher γ-GT activity (U/mg protein) in pellet and the lower supernatant-to-pellet ratio suggest that use of the microfuge yields a purer endothelial fraction.

Determine Vascular Space by Vascular Perfusion

At a given time after iv administration of the labeled protein and vascular marker albumin, blood is collected and the cerebral vascular contents are washed out with physiological buffer to remove the tracers remaining in the circulating blood. The brain-to-blood ratio of radioactivity in this group is compared with that in the group without vascular perfusion. Theoretically, albumin has no substantial uptake by the brain parenchyma, especially a few minutes after injection, but subtraction of the albumin values provides a further possible correction.

Tissue Processing

Tissue is homogenized with buffer in a glass homogenizer and mixed thoroughly with 26% dextran. This is centrifuged at 9000 g for 15–30 min at 4 °C in a swinging-bucket rotor. The pellet contains the vasculature and the supernatant contains the parenchymal/interstitial fluid. The ratio of the radioactivity of the protein in the supernatant or pellet to that in the serum is usually corrected by subtraction of the vascular space determined by the ratio of the labeled albumin.

This provides values for three compartments: (1) the microvascular compartment, representing vascular material and that tightly bound to the surface of the blood vessels as determined from the pellet in mice not undergoing washout; (2) reversible vascular compartment, representing material loosely adherent to the vessel walls or circulating cellular material, i.e., the difference in values between perfused and nonperfused animals; and (3) the parenchymal compartment, representing nonvascular material after washout. The presence of the injected protein in the parenchymal compartment indicates that it reached neuronal/glial cells and their processes or the interstitial fluid. This is a crucial measurement in the determination of whether the administered peptide/polypeptide was able to reach its desired target.

Stability

Choice of Method

In general, HPLC is most suitable *(33)* for peptides if care is taken to avoid shadowing *(34)*, but SDS-PAGE can be useful for polypeptides. No special modifications of these established techniques are necessary for BBB studies, although addition of a cocktail to inhibit cellular enzymatic activity liberated during processing of the brain is helpful. The use of a cocktail of enzyme inhibitors avoids the laborious effort of determination of the specific enzymes involved in degradation of each specific protein being analyzed, but may not be all-inclusive. The processing control provides an assessment of this requirement. Chromatographic analysis tends to underestimate the amount of injected protein reaching the brain since some degradation may have occurred after entry into the brain, even before homogenization. Acid precipitation, usually performed with 30% trichloroacetic acid containing oversaturated saline, usually correlates with the chromatographic methods, but cannot differentiate small peptide fragments produced by degradation.

Tissue Processing

Tissue is homogenized in phosphate-buffered saline with a glass homogenizer, to which the enzyme inhibitors are added. Enzyme inhibitors are not added to the blood since no homogenization is involved. However, since HPLC and SDS-PAGE are not performed in the cold, it also might be useful to inhibit any further enzymatic activity in the sample of serum. After centrifugation at 4 °C, the supernatant is removed, lyophilized, and then rehydrated shortly before chromatography. The elution position of the intact peptide is determined from the elution position of the stock solution. The percent of the total eluted radioactivity is calculated from the amount eluting at the same position as the standard. Correction can be made for processing by addition of the labeled peptide to blood and tissue obtained from a mouse processed in the same way but without *in vivo* injection.

A Note About Modeling

Crone's single-pass technique, introduced in the 1960s, analyzes the arterial-venous concentration difference of the test and reference substance *(35)*. The brain uptake index (BUI) method by Oldendorf *(36)* stresses the use of reference substances such as ^3H_2O and ^{14}C-butanol that have rapid, flow-dependent permeation to compensate for variations in cerebral blood flow. High BUI values are found with some individual amino acids, amines, hexoses, and substances such as nicotine, ethanol, caffeine, heroin (but not morphine), and cyanide *(1)*. The unidirectional flux in a rapidly equilibrating space is best illustrated by the Patlak–Blasberg–Fenstermacher treatment *(6–8)*. However, for slowly penetrating compounds, like small proteins, that are stable in the circulation for a short time, the multiple-time regression analysis coined by Banks and Kastin *(16,37)* is most appropriate.

Efflux After Intracerebroventricular (ICV) Administration

Background

If the administered peptide/polypeptide does not appear to enter the CNS, there are several possibilities, some of which were discussed previously in this review. First, of course, is that it really might not cross the BBB. This can be explained by such properties as low lipophilicity, high hydrogen bonding, self-aggregation, and unfavorable conformation. The injected protein also could be so rapidly degraded in blood that entry cannot be determined. It also could be tightly bound to serum proteins, in which case the influx would be much higher by in-situ perfusion in a blood-free buffer. Another possibility is that it exits the brain so rapidly that little remains in the brain at the time of experimental measurement of the injected material. Saturable efflux systems have been shown for a few peptides like Tyr-MIF-1, vasopressin, CRH, LHRH, and PACAP but have not been found for most of the other peptides we have tested *(37,38)*.

Procedure

The method described here is relatively simple, involving free-hand icv injection into the lateral ventricle of mice. In an anesthetized mouse, the first step is resection of a small amount of scalp to permit identification of the bregma (the intersection of the coronal and sagittal sutures). A hole is made in the skull 1 mm lateral and 0.2 mm posterior to the bregma with a 26-gauge hypodermic needle cuffed with polyethylene tubing so that only about 2.5 mm remains uncovered. This limits penetration to the roof of the ventricle. In the mice we use, this has been an effective site of injection, confirmed by autoradiography *(39)*. With a Hamilton syringe inserted to its premarked position 2.5 mm from the tip, 25,000 cpm in a volume of 1 µL are injected and the needle left in place for 5–10 s to reduce leakage. Dye can be used initially to verify localization, although the needle tract can be visualized without dye. Frequently used time points after injection are 2, 5, 10, and 20 min, individual mice being decapitated at each time. The 0-min value is determined in mice overdosed 10 min earlier with anesthesia to stop transport and bulk flow *(40)*, although other methods are described in the following section.

Calculation

The transport rate (*T*) is expressed in mol/g brain/min as calculated from the equation

$$T = (A - M)/Citw,$$

where *M* is the counts remaining in the brain, *C* the amount of protein injected expressed as moles, *i* the amount injected expressed in cpm, *t* the time in min from injection to decapitation, *w* the weight of the brain, and *A* the cpm of injected protein available. The value *A*, corresponding to the 0-min value, represents the cpm available for transport. Half-time disappearance is determined from the regression line of the log of brain radioactivity vs. time (0.301/slope).

Determination of *A* is most easily measured in mice overdosed 10 min before injection with anesthesia. This value also can be determined from the antilog of the point of intersection with the ordinate of the line between the log cpm remaining in brain and time. The values by either method are almost identical, as shown in a study of arginine vasopressin *(41)*. The efflux rate also can be determined from the rate of appearance of radioactivity in blood, but this is subject to peripheral influences like degradation *(40,41)*. Enzymatic activity in the CSF is negligible compared with that in blood, as shown with Tyr-MIF-1 *(42)*.

Altered permeation of the BBB after administration of an excess amount of the peptide, as also occurs with multiple-time regression analysis, provides strong physiological evidence for a saturable efflux transport system. Since use of excess unlabeled peptide/polypeptide to determine saturability after icv injection results in increased rather than decreased amounts remaining in the brain, this eliminates the possibility of leakage.

Statistical Comparisons Among Experimental Groups

When any group is used for more than one comparison, as with multiple t tests, the nominal probability levels no longer apply. The appropriate test is analysis of variance (ANOVA) followed by a multiple-range test. The test should be decided before the analysis. Among the various range tests, the order of increasing conservatism toward type I errors is Duncan, Newman–Keuls, Tukey, and Scheffé. Thus, the most stringent are most likely to avoid type I errors but have the inherent disadvantage of being less sensitive in detecting real differences, so that they run the risk of making an increased number of type II errors. The Dunnett test is useful for comparisons between a single control group and several experimental groups. Most statisticians believe that multiple comparison tests are only valid if the ANOVA is significant, but not all agree on that, although they do agree that more attention should be given to significant interactions. For multiple-time regression analysis, two means (slope and intercept) are calculated from the data, so that $n - 1$ is used as the value for sample size.

Conclusion

The techniques described in this review facilitate quantification of the influx and efflux of small proteins across the BBB. Hopefully they will stimulate useful studies to advance understanding of the communication between the brain and body as well as therapy of CNS disorders.

References

1. Oldendorf, W. Brain uptake of radiolabeled amino acids, amines, and hexoses after arterial injection. *Am. J. Physiol.*, 1971; **221**: 16929–16939.
2. Rapoport, S.I. *Blood-Brain Barrier in Physiology and Medicine*. Raven Press, New York, 1976.

3. Davson, H., and Segal, M.B. *Physiology of the CSF and Blood–Brain Barriers.* CRC Press, Boca Raton, FL, 1996.

4. Begley, D.J. Peptides and the blood-brain barrier. In *Handbook of Experimental Pharmacology, Physiology and Pharmacology of the Blood–Brain Barrier,* Vol. 103, M.W.B. Bradbury, ed. Springler-Verlag, Berlin, 1992, pp. 151–203.

5. Kastin, A.J., Hahn, K., and Zadina, J.E. Regional differences in peptide degradation by rat cerebral microvessels: A novel regulatory mechanism for communication between blood and brain. *Life Sci.,* 2001; **69**: 1305–1312.

6. Fenstermacher, J.D. Methods for quantifying the transport of drugs across blood–brain barrier systems. *Pharmacol. Ther.,* 1981; **14**: 217–248.

7. Blasberg, R.G., Fenstermacher, J.D., and Patlak, C.S. Transport of α-aminoisobutyric acid across brain capillary and cellular membranes. *J. Cereb. Blood Flow Metab.,* 1983; **3**: 8–22.

8. Patlak, C.S., Blasberg, R.G., and Fenstermacher, J.D. Graphical evaluation of blood-to-brain transfer constants from multiple time uptake data. *J. Cereb. Blood Flow Metab.,* 1983; **3**: 1–7.

9. Kastin, A.J., Nissen, C., and Coy, D.H. Permeability of the blood–brain barrier to DSIP peptides. *Pharmacol. Biochem. Behav.,* 1981; **15**: 955–959.

10. Banks, W.A., Jaspan, J.B., and Kastin, A.J. Selective, physiological transport of insulin across the blood-brain barrier: Novel demonstration by species-specific radioimmunoassays. *Peptides,* **1997;** **18**: 1257–1262.

11. Hunter, W.M., and Greenwood, F.C. Preparation of iodine-131 labelled human growth hormone of high specific activity. *Nature,* 1962; **194**: 495–496.

12. Marchalonis, J. An enzymic method for the trace iodination of immunoglobulins and other proteins. *Biochem. J.,* 1969; **113**: 299–305.

13. Bolton, A.E., and Hunter, W.M. A new method for labelling protein hormones with radioiodine for use in the radioimmunoassay. *J. Endocrinol.,* 1972; **55**: xxx–xxxi.

14. O'Rourke. E.C., Drummond, R.J., and Creasey, A.A. Binding of ^{125}I-labeled recombinant βinterferon (IFN-β Ser$_{17}$) to human cells. *Mol. Cell. Biol.,* 2006; 4: 2745–2749.

15. Banks, W.A., and Kastin, A.J.. Opposite direction of transport across the blood–brain barrier for Tyr-MIF-1 and MIF-1: Comparison with morphine. *Peptides,* 1994; **15**: 23–29.

16. Kastin, A.J., Akerstrom, V., and Pan, W. Validity of multiple-time regression analysis in measurement of tritiated and iodinated leptin crossing the blood–brain barrier: Meaningful controls. *Peptides,* 2001; **22**: 2127–2136.

17. Banks, W.A., and Kastin, A.J. Peptides and the blood–brain barrier: Lipophilicity as a predictor of permeability. *Brain Res. Bull.,* 1985; **15**: 287–292.

18. Chikhale, E.G., Ng, K.Y., Burton, P.S., and Borchardt, R.T. Hydrogen bonding potential as a determinant of the *in vitro* and *in situ* blood–brain barrier permeability of peptides. *Pharm. Res.,* 1994; **11**: 412–419.

19. Egleton, R.D., and Davis, T.P. Bioavailability and transport of peptides and peptide drugs into the brain. *Peptides,* 1997; **18**: 1431–1439.

20. Kastin, A.J., Pan, W., Maness, L.M., and Banks, W.A. Peptides crossing the blood–brain barrier: Some unusual observations. *Brain Res.,* 1999; **848**: 96–100.

21. Pan, W., Banks, W.A., and Kastin, A.J. Permeability of the blood–brain and blood-spinal cord barriers to interferons. *J. Neuroimmunol.,* 1997; **76**: 105–111.

22. Banks, W.A., and Kastin, A.J. Differential permeability of the blood–brain barrier to two pancreatic peptides: Insulin and amylin. *Peptides*, 1998; **19**: 883–889.

23. Nonaka, N., Banks, W.A., Mizushima, H., Shioda, S., and Morley, J.E. Regional differences in PACAP transport across the blood-brain barrier in mice: A possible influence of strain, amyloid beta protein, and age. *Peptides*, 2002; **12**: 2197–2202.

24. Pan, W., Ding, Y., Yu, Y., Ohtaki, H., Nakamichi, T., and Kastin, A.J. Stroke upregulates TNF alpha transport across the blood–brain barrier. *Exp. Neurol.*, 2006; **198**: 222–233.

25. Kastin, A.J., and Pan, W. Editorial: Intranasal leptin: Blood–brain barrier bypass (BBBB) for obesity? *Endocrinology*, 2006; **147**: 2086–2087.

26. Johanson, C.E. The choroid plexus–CNF nexus. In *Neuroscience in Medicine*, P.M. Conn, ed. Humana Press Inc., Totawa, Press, 2003, pp. 165–195.

27. Plotkin, S.R., Banks, W.A., and Kastin, A.J. Comparison of saturable transport and extracellular pathways in the passage of interleukin-1 αacross the blood–brain barrier. *J. Neuroimmunol.*, 1996; **67**: 41–47.

28. Maness, L.M., Kastin, A.J., and Banks, W.A. Relative contributions of a CVO and the microvascular bed to delivery of blood-borne IL-1α to the brain. *Am. J. Physiol.*, 1998; **275**: E207–E212.

29. Triguero. D., Buciak, J., and Pardridge, W.M. Capillary depletion method for quantification of blood–brain barrier transport of circulating peptides and plasma proteins. *J. Neurochem.*, 1990; **54**: 1882–1888.

30. Gutierrez, E.G., Banks, W.A., and Kastin, A.J. Murine tumor necrosis factor alpha is transported from blood to brain in the mouse. *J. Neuroimmunol.*, 1993; **47**: 169–176.

31. Pan, W., Vallance, K.L., and Kastin, A.J. TGF alpha and the blood–brain barrier: Accumulation in cerebral vasculature. *Exp. Neurol.*, 1999; **160**: 454–459.

32. Yu, C., Kastin, A.J., Ding, Y., and Pan, W. Gamma glutamyl transpeptidase is a dynamic indicator of endothelial response to stroke. *Exp. Neurol.*, 2007; **203**: 116–122.

33. Hoke, F. Recent advances increase HPLC use in life sciences. *Scientist*, 1993; **7**: 18–19.

34. Fischman, A.J., Kastin, A.J., and Graf, M.V. HPLC shadowing: Artifacts in peptide characterization monitored by RIA. *Peptides*, 1984; **5**: 1007–1010.

35. Crone, C. The permeability of capillaries of various organs as determined by the use of the "indicator diffusion" method. *Acta Physiol. Scand.*, 1963; **58**: 292–305.

36. Olendorf, W.H. Measurement of brain uptake of radiolabeled substances using a tritiated water internal standard. *Brain Res.*, 1970; **24**: 372–376.

37. Kastin, A.J., Pan, W. Piptide transport across the blood-brain barrier. In Prokai, L., Prokai-Tatrai, K. eds. Piptide Transport and Delivery into the Central Nervous System, pp. 79–100. Basel, Switzerland: Birkhauser Verlag, 2003.

38. Pan, W., Kastin, A.J. Why study transport of peptides and proteins at the neurovascular interface. *Brain Res. Rev.*, 2004; **46**: 32–43.

39. Maness, L.M., Kastin, A.J., Farrell, C.L., and Banks, W.A. Fate of leptin after intracerebroventricular injection into the mouse brain. *Endocrinology*, 1998; **139**(11): 4556–4562.

40. Banks, W.A., and Kastin, A.J. Quantifying carrier-mediated transport of peptides from the brain to the blood. In *Methods in Enzymology*, Vol. 168, P.M. Conn, ed. Academic Press, San Diego, 1998; pp. 652–660.
41. Banks, W.A., Kastin, A.J., Horvath, A., and Michals, E.A. Carrier-mediated transport of vasopressin across the blood-brain barrier of the mouse. *J. Neurosci. Res.*, 1987; **18**: 326–332.
42. Kastin, A.J., Banks, W.A., Hahn, K., and Zadina, J.E. Extreme stability of Tyr-MIF-1 in CSF. *Neurosci. Lett.*, 1994; **174**: 26–28.

9

Cancer Cell Receptor Internalization and Proliferation: Effects of Neuropeptide Analogs

Terry W. Moody, Michael Schumann, and Robert T. Jensen

Abstract

Neuropeptide receptors can be used as molecular targets to deliver chemotherapeutic drugs into cancer cells. The gastrin-releasing peptide (GRP) receptor was used to deliver a camptothecin (CPT)–bombesin (BB) conjugate into the lung cancer cell line NCI-H1299. The CPT–BB conjugate was metabolized by NCI-H1299 intracellular enzymes releasing the cytotoxic CPT. Receptor binding methods and fluorescent methods are presented for studying the internalization of GRP receptors. The MTT, clonogenic, and ^{3}H-thymidine uptake assays are presented for studying the proliferation of NCI-H1299 cells. The results indicate that CPT-BB conjugates are internalized as a result of receptor-mediated endocytosis.

Introduction

Neuropeptides of the bombesin (BB) family function as modulators in the central nervous system (CNS). BB-like peptides are present in discrete brain regions such as the paraventricular nucleus of the hypothalamus as well as the nucleus tractus solitarius *(1)*. When released from presynaptic neurons, they bind to receptors in postsynaptic cells *(2,3)*. BB administration in the rat CNS results in decreased food intake *(4)*. This tetradecapeptide was initially isolated from skin of the frog *Bombina bombina (5)*, but injection into rat brain ventricles caused hypothermia and hyperglycemia *(6,7)*. Also, BB injection resulted in increased rat grooming and gastric acid secretion *(8,9)*. BB-like peptides function in a paracrine manner in the CNS to alter rodent behavior *(10)*.

From: *Neuromethods, Vol. 39: Neuropeptide Techniques*
Edited by: I. Gozes © Humana Press Inc., Totowa, NJ

In human small-cell lung cancer (SCLC), BB-like peptides function as autocrine growth factors *(11)*. The secretion rate of BB-like peptides from small-cell lung cancer cells is increased after the addition of VIP, which increases the intracellular cAMP *(12)*. The BB-like peptides bind to cell surface gastrin-releasing peptide (GRP) receptors and the peptide-receptor complex is internalized *(13)*. Addition of BB to SCLC cells causes phosphatidylinositol (PI) turnover, leading to the activation of MAP kinase and increased proliferation *(14,15,16)*. The actions of BB are reversed by GRP receptor antagonists such as (Psi[13,14], Leu[14])BB, BW-2258U89, or PD176252 *(17,18)*. BW2258U89 or PD176252 functions as cytostatic agents and inhibits the growth of SCLC *in vitro* and *in vivo* *(19)*.

BB is structurally related to the human peptides gastrin-releasing peptide (GRP) and neuromedin B (NMB), which contain 27 and 10 amino acids, respectively *(20,21)*. Table 1 shows that the C-terminal of BB, GRP, and NMB has sequence homologies. While BB and GRP bind with high affinity to the GRP receptor, NMB binds with high affinity to the NMB receptor *(22,23)*. The GRP and NMB receptors contain 384 and 390 amino acids, respectively, and have approximately 50% sequence homology *(24,25)*. In contrast, the orphan receptor bombesin receptor subtype (BRS)-3 contains 399 amino acids and binds GRP as well as NMB with low affinity *(26,27,28)*. Peptide agonists have been developed that bind with high affinity to all classes of BB receptors *(29,30)*, and BB receptors have been detected in numerous human cancer biopsy specimens as well as in cell lines *(31,32,33)*.

Recently, chemotherapeutic-peptide conjugates have been developed that kill cancer cells *(34,35)*. Camptothecin (CPT)–BB conjugates are cytotoxic for SCLC cells *(36)* as well as the NSCLC cell line NCI-H1299 *(37)*. The CPT–BB conjugates, such as CPT-L2-BA3, contain a carbamate linker (L2) as well as a BB analog (BA3), which binds with high affinity to GRP and NMB receptors as well as BRS-3. CPT-L2-BA3 binds with high affinity to GRP receptors present on NCI-H1299 cells. CPT-L2-BA3 functions as a GRP receptor agonist increasing NCI-H1299 PI turnover and cytosolic Ca^{2+} concentrations *(36)*. The prodrug CPT-L2-BA3-GRP receptor complex is internalized in clathrin-coated pits; however, in acidic endosomes, the complex dissociates. The GRP receptor may be recycled, whereas the CPT-L2-BA3 is metabolized in lysosomes. The drug CPT is subsequently released, diffuses into the nucleus, and inhibits topoisomerase 1 activity, leading to cancer cell apoptosis *(38)*. Thus, the CPT–BB conjugate is cytotoxic for NCI-H1299 cells. This chapter focuses on the methods for studying the internalization of peptide-GRP receptors and proliferation assays.

Internalization

Cell Culture

The internalization of neuropeptides by lung cancer cells was demonstrated by receptor binding techniques. NCI-H1299 cells were cultured in Roswell Park Memorial Institute (RPMI-1640) medium containing 10% heat-inactivated fetal bovine serum (FBS) (Life Technologies Inc., Grand Island, NY). NCI-H1299 cells were split weekly 1/20 with trypsin-EDTA. A T175 flask with a monolayer of cells contains 40×10^6 cells and a new flask is seeded with 2×10^6 cells in 40 mL of RPMI-1640 with 10% FBS. Routinely cells were split with trypsin-EGTA (Life Technologies Inc.) on Mondays and fed on Thursdays. HEK 293 cells were cultured in DMEM medium, supplemented with 10% FBS, 100 U/mL penicillin, and 100 mg/mL streptomycin. The cells [NCI-H1299 (human NSCLC) and HEK 293 (human embryonic kidney)] were from American Type Culture Collection (ATCC) (Rockville, MD), were mycoplasma-free, and were used when they were in exponential growth phase after incubation at 37 °C in 5% CO_2, 95% air.

Receptor Binding Studies

Neuropeptides were labeled with ^{125}I and used in radio-receptor assays. CPT-L2-BA3 was iodinated with 0.8 μg of IODO-GEN (Pierce Chemical Co., Rockford, IL) solution (0.01 μg/μL in chloroform), which was added to a 5-mL plastic test tube, dried under nitrogen, and washed with 100 mL of 0.5 M potassium phosphate solution (pH 7.4). To this tube were added 20 μL of potassium phosphate of the appropriate pH, 8 μg of CPT-L2-BA3 in 4 μL of water, and 2 mCi (20 μl) of $Na^{125}I$ and incubated for 6 min at room temperature [$Na^{125}I$ (2200 Ci/mmol) was from Amersham Pharmacia Biotech]. The incubation was stopped with 300 μL of water. ^{125}I-CPT-L2-BA3 was separated using a C18 Sep-Pak (Waters Associates, Milford, MA) and further purified by reverse-phase, high-performance liquid chromatography on a C18 column. The fractions with the highest radioactivity and binding were neutralized with 0.2 M Tris buffer (pH 9.5) and stored with 0.5% bovine serum albumin (w/v) at −20 °C.

NCI-H1299 cells (1×10^5/well) were placed into 24-well plates containing 1 mL of RPMI-1640 containing 10% FBS. After 2 days, the cells were incubated with 100,000 cpm of ^{125}I-CPT-L2-BA3 (0.25 nM) at 37 °C for 30 min in 250 µL of receptor binding buffer containing PBS with 0.25% bovine serum albumin (BSA) and 0.25 mg/mL bacitracin (Sigma Chemical Co., St. Louis, MO). The supernatant was removed and the 24-well plates rinsed three times with 0.25 mL of receptor binding buffer at 4 °C. The cells that contained bound radiolabeled peptide were solubilized with 0.25 mL of 0.2 N NaOH at 25 °C for 30 min. The samples were counted in an LKB gamma counter. The nonsaturable binding was the amount of radioactivity associated with cells in incubations containing 0.25 nM of radioligand (2200 Ci/mmol) and 1 µM of unlabeled CPT-L2-BA3. Nonsaturable binding was <10% of total binding in all the experiments. Dissociation constants (K_i) were determined using a least-squares curve-fitting program (KaleidaGraph) and the Cheng–Prusoff equation *(39)*.

Peptide Internalization

NCI-H1299 cells were pre-incubated with isotonic sucrose or 25 µM chlorpromazine at 25 °C for 30 min. Then ^{125}I-CPT-L2-BA3 was added in the presence or absence of competitor at 37 °C for the indicated period of time. The plates were rinsed three times in receptor binding buffer at 4 °C. Radiolabeled ^{125}I-CPT-L2-BA3 bound to the cell surface was removed by treatment with 0.25 mL of 0.2 M acetic acid (pH 2.5) and 0.5 M NaCl at 4 °C. After 5 min, the supernatent was removed and the cells, which contained internalized ^{125}I-CPT-L2-BA3, were solubilized in 0.25 mL of 0.2 N NaOH and counted in the gamma counter.

^{125}I-CPT-L2-BA3 specifically bound to NCI-H1299 NSCLC cells and CPT-L2-BA3 (IC_{50} = 0.8 ± 0.2 nM) had a fivefold greater affinity for hGRP receptors on these cells than BA0 (IC_{50} = 4.3 ± 0.4 nM) (Fig. 1, left panel). These results indicate that CPT was coupled to the N-terminal of a BB analog, with retention of high-affinity binding activity. Surprisingly, CPT-L2-BA3 bound with higher affinity than did BA0, from which it was derived (Table 1). This may result because the L2-BA3 binds with high affinity to

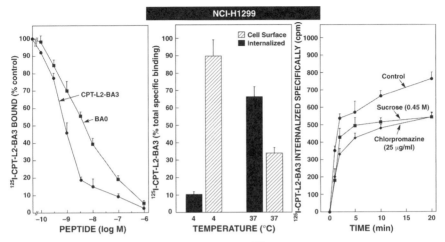

Fig. 1. Binding and internalization of ^{125}I-CPT-L2-BA3 by NCI-H1299 cells. (Left) The % specific ^{125}I-CPT-L2-BA3 bound was determined as a function of unlabeled CPT-L2-BA3 (●) or BA0 (■) using NCI-H1299 cells. (Middle) The percentage of specifically bound ^{125}I-CPT-L2-BA3 bound to the cell surface (▨) or internalized (■) by NCI-H1299 cells was determined at 4 °C or 37 °C. (Right) The cpm of specifically internalized ^{125}I-CPT-L2-BA3 to NCI-H1299 cells was determined at 37 °C as a function of time after no additions (●), addition of chlorpromazine [25 (μg/mL (▲)], or addition of 0.45 M sucrose (■). The mean value ± S.D. of four determinations is indicated. This experiment is representative of two others.

Table 1. Structure of BB-like Peptides

BB	Pyr-Gln-Arg-Leu-Gly-Asn-Gln-Trp-Ala-Val-Gly-His-Leu-Met-NH$_2$
GRP	Ala-Pro-Val-Ser-Val-Gly-Gly-Gly-Thr-Val-Leu-Ala-Lys-Met-Tyr-Pro-Arg-Gly-Asn-His-Trp-Ala-Val-Gly-His-Leu-Met-NH$_2$
NMB	Gly-Asn-Leu-Trp-Ala-Thr-Gly His-Phe-Met-NH$_2$
BA0	DTyr-Gln-Trp-Ala-Val-βAla-His-Phe-Nle-NH$_2$
CPT-L2-BA3	CPT-Nmethyl-aminoethyl-Gly-DSer-DTyr-Gln-Trp-Ala-Val-βAla-His-Phe-Nle-NH$_2$

[a] Sequence homologies relative to BB are underlined.

the binding site on the GRP receptor, but in addition the CPT interacts with an additional hydrophobic site on the GRP receptor. Similarly, CPT-L2-BA3 bound with high affinity to cells transfected with NMB receptors as well as BRS-3 (data not shown).

^{125}I-CPT-L2-BA3 was rapidly internalized by NCI-H1299 NSCLC cells (Fig. 1, middle panel). Specifically, at 4 °C, only 10% of the ^{125}I-CPT-L2-BA3 was internalized, whereas after a 5-min incubation at

37 °C with the NCI-H1299 cells, 67% of the specifically bound [125] I-CPT-L2-BA3 was internalized (Fig. 1, middle panel). These results indicate that the ^{125}I-CPT-L2-BA3-GRP receptor complex is rapidly internalized. Similarly, ^{125}I-CPT-L2-BA3 was rapidly internalized in cells transfected with NMB receptors as well as BRS-3 (data not shown).

Agonist-occupied G protein receptors are frequently internalized via a clathrin-coated pit mechanism *(40)*. Previously it was shown that the GRP receptor is internalized by peptide agonists *(41,42)*. To investigate whether ^{125}I-CPT-L2-BA3 was internalized by a clathrin-dependent mechanism in NCI-H1299 NSCLC cells, the effects on its internalization by hypertonic sucrose *(43)*, as well as chlorpromazine *(44)*, two specific inhibitors of clathrin-coated endocytosis, were determined (Fig. 1 , right panel). Both 0.45 M sucrose and chlorpromazine (25 µg/mL) inhibited ^{125}I-CPT-L2-BA3 internalization by 35% (Fig. 1 , right panel), demonstrating that a clathrin-coated pit mechanism was at least partially involved in its internalization by these cells. Often receptors associated with clathrin-coated pits are recycled to the cell surface, whereas the neuropeptides are degraded by lysozomal enzymes. Preliminary data (R. Jensen, unpublished) indicate that ^{125}I-L2-BA3 is metabolized to CPT and ^{125}I-L2-BA3 after exposure to cells transfected with human GRP receptors.

Fluorescent Microscopy

Internalization of ligands and receptors was demonstrated using fluorescent microscopic techniques. HEK 293 cells were seeded on 6-well plates at a density of 0.5×10^6 cells per well. A sequence encoding the HA epitope tag (YPYDVPDYA) was inserted between the first (Met) and second (Ala) amino acid residues of the GRPR using the ExSite PCR-Based Site-Directed Mutagenesis Kit (STRATAGENE, La Jolla, CA), following the manufacturer's instruction. The mammalian expression vector, pcDNA3, custom primers, and restriction endonucleases (BamHI, HindIII, XbaI, and EcoRI) were from Invitrogen (Carlsbad, CA). Nucleotide sequence analysis of the entire coding region was performed using an automated DNA sequencer (ABI Prism 377 DNA sequencer; Applied Biosystems Inc., Foster City, CA). Cells were transfected using 3 µL of Fugene 6 reagent and 1 µg of HA-GRPR-pcDNA3 following the manufacturer's protocol. One day after transfection, the cells were trypsinized and plated on poly-Lysine-coated glass coverslips in 12-wells at a density of 100,000 cells/well. Two days later, cells were washed with PBS and treated with or without the CPT-L2-BA3 (3 nM) in DMEM for various incubation times at 37 °C. The cells were washed three times in buffer to remove free CPT-L2-BA3. The cells containing bound CPT-L2-BA3 were fixed (paraformaldehyde 4%, 10 min, 22 °C), permeabilized (Triton-X-100 0.5%, 5 min, 22 °C), and blocked (2% BSA, 2% donkey serum in PBS, 30 min, 22 °C). Cells were stained with a polyclonal rabbit anti-bombesin antibody (1/500 dilution) as the first antibody and an FITC-labeled donkey anti-rabbit antibody antibody (1/200 dilution) as the secondary antibody. The HA-GRPR was stained using a mouse anti-HA antibody (1/100 dilution; Santa Cruz, Santa Cruz, CA) as the first and a TRITC-labeled donkey anti-mouse antibody (1/200 dilution) as the secondary antibody (ImmunoResearch Laboratories, West Grove, PA). Nuclei were visualized after DAPI counterstaining. Coverslips were mounted using Vectorshield (Vector, Burlingame, CA) and fixed on glass slides with nail polish. Imaging was done using a Nikon fluorescent microscope.

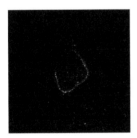  

Fig. 2. Internalization of the CPT-L2-BA3 in HEK 293 cells transiently expressing HA-tagged GRP receptors. HEK 293 cells (0.5×10^6 cells/6-well plate) were transfected with 1 μg of HA-GRPR-pcDNA3 and split to poly-Lysine-coated 12-wells the following day. Three days after transfection, cells were treated with or without CPT-L2-BA3 (3 nM) for 20 min. After removal of free CPT-L2-BA3, cells containing bound CPT-L2-BA3 were fixed, permeabilized, and stained with a polyclonal bombesin antibody (left) or an HA antibody against the receptor (middle), as described in the section on materials and methods. Merge images of the pictures (right) are shown to reveal co-localization of the BB analog and the receptor in the plasma membrane but not intracellularly. Shown are representative pictures of three independent experiments.

Using HEK cells transfected with an epitope-tagged GRPR (HA-GRPR), the receptor was localized (red color) to the plasma membrane with no agonist present. Twenty minutes after CPT-L2-BA3 (10 nM) was added, both the GRP receptors (red color) (Fig. 2 , middle) and BB-like immunoreactivity (green color) (Fig. 2 ,left) were localized to both the plasma membrane and intracellularly. On the plasma membrane, they were frequently co-localized (yellow color) (Fig. 2 , right). Some of the GRP receptors (red color) had already internalized after 20 min (Fig. 2 , middle). Similarly, some of the neuropeptide (green color) was internalized (Fig. 2, right); however, there was little intracellular co-localization. These results suggest that CPT-L2-BA3 and the GRP receptor co-localize at the plasma membrane, but after internalization, the bound CPT-L2-BA3 dissociates from the GRP receptor. Also, Fig. 2, right, shows that many of the cells that counterstained blue lacked GRP receptors and BB-like immunoreactivity.

Proliferation Assays

MTT Assay

If CPT is released after intracellular metabolism of CPT-L2-BA3, it should be cytotoxic for NCI-H1299 cells. Growth studies *in vitro* were conducted using the [3-(4,5 dimethylthiazol-2-yl)-2.5-diphenyl-2H-tetrazolium bromide] (MTT) (Sigma Chemical Co., St. Louis, MO) colorimetric assays. NCI-H1299 cells (10^4/well) were placed in SIT medium (100 mL, RPMI-1640 containing 3×10^{-8} M Se_2O_3, 5 ug/mL insulin, and 10 ug/mL transferrin), and various concentrations of CPT-L2-BA3 were added. After 3 days, 15 μL (1 mg/mL) of MTT were added. After 4 h, 150 μL of DMSO were added. After 16 h, the optical density at 570 nm was determined using an ELISA reader. Using the MTT assay, CPT-L2-BA3 caused a dose-dependent inhibition of proliferation by NCI-H1299 cells (Fig. 3). CPT-L2-BA3 had a half-maximal (IC_{50}) inhibitory growth effect in NCI-H1299 NSCLC cells of 190 ± 20 nM. In contrast, BA0 had little effect in the MTT assay, whereas CPT strongly killed NCI-H1299 cells (T. Moody, unpublished).

Clonogenic Assay

The MTT assay is used to evaluate cytotoxic agents, whereas the clonogenic assay is used to evaluate cytotoxic as well as cytostatic

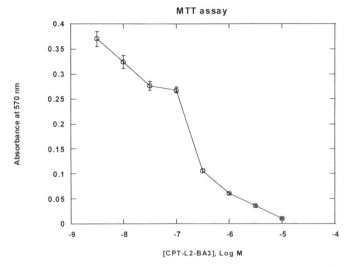

Fig. 3. MTT assay. NCI-H1299 cells were plated into 96-well plates in the presence of varying concentrations of CPT-L2-BA3. After 3 days, MTT was added. After 4 h, DMSO was added and the absorbance determined at 570 nm. The mean value ± S.D. of eight determinations is indicated. This experiment is representative of four others.

agents. The base layer consists of 2 mL of 0.5% agarose in SIT medium containing 5% fetal bovine serum in 6-well plates. The agar is boiled for 1 h in SIT medium, then poured at 37 °C, and allowed to cool at room temperature (22 °C) and solidify. The top layer consists of 2 mL of SIT medium in 0.3% agarose (FMC Corp., Rockford, ME), CPT-BB conjugates, and 5×10^4 NCI-H1299 cells. Triplicate wells were plated; after 2 weeks, 1 mL of 0.1% p-iodonitrotetrazolium violet was added. After 16 h at 37 °C, the plates were screened for colony formation; the number of colonies larger than 50 μm in diameter was counted using an Omnicon image analysis system.

Figure 4 shows that BA0 had little effect on colony formation. In contrast, CPT-L2-BA3 inhibited all colony formation at 1000 nM, whereas colony formation was half-maximally inhibited at 300 nM. Recently, we found that the highly fluorescent CPT is released by the weakly fluorescent CPT-L2-BA3, 2 days but not 1 h after exposure to NCI-H1299 cells (T. Moody, unpublished). These results suggest that it takes a period of time for CPT to concentrate in the nucleus of NCI-H1299 cells and inhibit topoisomerase-1.

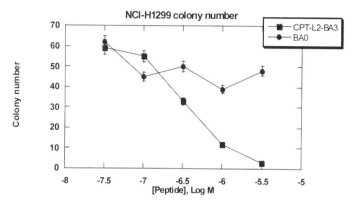

Fig. 4. Clonogenic assay. The ability of varying concentrations of CPT-L2-BA3 (■) and BA0 (●) to inhibit the clonal growth of NCI-H1299 cells was determined. Each assay was performed in triplicate and the mean value ± S.D. of three experiments is indicated.

Amino Acid and Nucleic Acid Uptake

The ability of CPT-L2-BA3 to alter DNA replication and protein synthesis can be investigated using radioactive isotopes. NCI-H1299 cells in 24-well plates were incubated with various concentrations of CPT-L2-BA3 in RPMI-1640 containing 10% FBS.

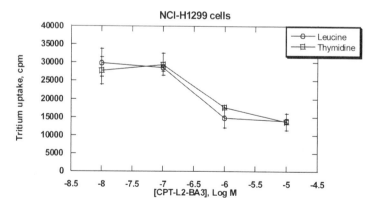

Fig. 5. Amino acid and nucleic acid uptake. The cpm of ^{3}H-leucine (○) and ^{3}H-thymidine (□) taken up by NCI-H1299 cells was determined as a function of CPT-L2-BA3 concentration. The mean value ± S.D. of four determinations is indicated. This experiment is representative of two others.

After 3 days, ^{3}H-thymidine (10^6 cpm) or ^{3}H-leucine (10^6 cpm) was added for 2 h. Free radiolabel was removed and the cells washed three times with cold PBS. The cells were lysed with 250 uL of 0.2 N HCl followed by 250 uL of 0.2 N NaOH. The samples were put in a scintillation vial, and 10 mL of Aquasol were added. The samples were counted in a β-counter. Figure 5 shows that 10 and 100 nM of CPT-L2-BA3 had little effect on ^{3}H-thymidine or ^{3}H-leucine incorporation. In contrast, 1000 and 10000 nM of CPT-L2-BA3 strongly inhibited ^{3}H-thymidine or ^{3}H-leucine incorporation. When CPT inhibits topoisomerase 1, DNA cannot unwind and be replicated. If DNA synthesis is sufficiently inhibited, mRNA transcription and protein translation will also decrease. Figure 5 shows that ^{3}H-leucine incorporation is strongly decreased using 1 uM of CPT-L2-BA3.

Conclusions

A goal in translational research is to develop new modes of drug therapy using molecular targets. Here the GRP receptor was used as a molecular target to deliver a prodrug CPT-L2-BA3 to lung cancer cells. CPT-L2-BA3 was hydrophilic and readily soluble in water, whereas CPT was hydrophobic and insoluble in water. More hydrophilic analogs of CPT such as irinotecan or topotecan inhibit topoisomerase 1 are used to treat lung cancer patients, but there are toxic side effects such as neutropenia *(45)*.

Because GRP receptors are internalized after binding peptide agonists such as CPT-L2-BA3, internalization methods were presented. Once internalized, CPT-L2-BA3 is metabolized, releasing the drug CPT, which kills cancer cells. Thus, methods to evaluate proliferation of cancer cells were presented. These assays are applicable to neuropeptides outside the BB family. We have found that cholecystokinin receptor antagonists, neurotensin receptor antagonists, somatostatin receptor agonists, and vasoactive intestinal peptide receptor antagonists inhibit SCLC cellular proliferation *(33)*. Somatostatin receptor agonists but not cholecystokinin receptor, GRP receptor, neurotensin receptor, or vasoactive intestinal peptide receptor antagonists are internalized. Recently, a CPT–somatostatin conjugate was described that is cytotoxic for lung cancer cells *(38)*. These results indicate that neuropeptide receptors can serve as molecular targets to deliver cytotoxic agents to cancer cells.

Acknowledgments

The authors thank Drs. David Coy, Joseph Fusilier, Samuel Mantey, Alfredo Martinez, Tomoo Nakagawa, and Tapas Pradhan.

References

1. Moody, T.W., O'Donohue, T.L., and Jacobowitz, D.M. Biochemical localization and characterization of bombesin-like peptides in discrete regions of rat brain. *Peptides,* 1981; **2:** 75–79.
2. Moody, T.W., Pert, C.B., Rivier, J., and Brown, M.R. Bombesin: Specific binding to rat brain membranes. *Proc. Natl. Acad. Sci. USA,* 1978; **75:** 5372–5376.
3. Moody, T.W., Thoa, N.B., O'Donohue, T.L., and Pert, C.B. Bombesin-like peptides in rat brain: Localization in synaptosomes and release from hypothalamic slices. *Life Sci.,* 1980; **26:** 1707–1712.
4. Gibbs, J., Fauser, D.J., Rowe, E.A., Rolls, B.J., Rolls, E.T., and Maddison, S.P. Bombesin suppresses feeding in rats. *Nature,* 1979; **282:** 208–210.
5. Anastasi, A., Erspamer, V., and Bucci, M. Isolation and amino acid sequences of alytesin and bombesin: Two analogous active tetradecapeptides from the skin of European discoglossid frogs. *Arch. Biochem. Biophys.,* 1973; **148:** 443–446.
6. Brown, M.R., Rivier, J., and Vale, W. Bombesin: Potent effects on thermoregulation in the rat. *Science,* 1977; **196:** 998–1000.
7. Brown, M., Rivier, J., and Vale, W. Bombesin affects the central nervous system to produce hyperglycemia in rats. *Life Sci.,* 1978; **21:** 1729–1734.
8. Merali, Z., Moody, T., Kateb, P., and Piggins, H. Antagonism of satiety and grooming effects of bombesin by antiserum to bombesin and by [Tyr[4], D-Phe[12]]bombesin: Central versus peripheral effects. *Ann. NY Acad. Sci.,* 1988; **547**: 489–492.
9. Tache, Y. CNS peptides and regulation of gastric acid secretion. *Ann. Rev. Physiol.,* 1988; **50**: 19–39.
10. Merali, Z., Kent, P., and Anisman, H. Role of bombesin-related peptides in the mediation or integration of the stress response. *Life Sci.,* 2002; **59**: 272–287.
11. Cuttitta, F., Carney, D.N., Mulshine, J., Moody, T.W., Fedorko, J., Fischler, A., and Minna, J.D. Bombesin-like peptides can function as autocrine growth factors in human small cell lung cancer. *Nature,* 1985; **316:** 823–825.
12. Korman, L.Y., Carney, D.N., Citron, M.L., and Moody, T.W. Secretin/VIP stimulated secretion of bombesin-like peptides from human small cell lung cancer. *Cancer Res.,* 1986; **46:** 1214–1218.
13. Moody, T.W., Mahmoud, S., Staley, J., Naldini, L., Cirillo, D., South, V., Felder, S., and Kris, R. Human glioblastoma cell lines have neuropeptide receptors for bombesin/GRP. *J. Mol. Neurosci.,* 1989; **1**: 235–242.
14. Benya, R.V., Kusui, T., Pradhan, T.K., Battey, J.F., and Jensen, R.T. Expression and characterization of cloned human bombesin receptors. *Mol. Pharmacol.,* 1995; **47:** 10–20.
15. Carney, D.N., Cuttitta, F., Moody, T.W., and Minna, J.D. Selective stimulation of small cell lung cancer clonal growth by bombesin and gastrin releasing peptide. *Cancer Res.,* 1987; **47:** 821–825.

16. Koh, S.W., Leyton, J., and Moody, T.W. Bombesin activates MAP kinase in non-small cell lung cancer cells. *Peptides*, 1999; **20**: 121–126.
17. Mahmoud, S., Staley, J., Taylor, J., Bogden, A., Moreau, J.P., Coy, D., Avis, I., Cuttitta, F., Mulshine, J.L., and Moody, T.W. [Psi [13,14]] bombesin analogues inhibit growth of small cell lung cancer *in vitro* and *in vivo*. *Cancer Res.*, 1991; **51**: 1798–1802.
18. Moody, T.W., Zia, F., Venugopal, R., Patierno, S., LeBan, J., and McDermod, J. BW2258: A GRP receptor antagonist which inhibits small cell lung cancer growth. *Life Sci.*,1995; **56**: 523–529.
19. Moody, T.W., Leyton, J., Garcia, L., and Jensen, R.T. Nonpeptide gastrin releasing peptide receptor antagonists inhibit the proliferation of lung cancer cells. *Eur. J. Pharmacol.*, 2003b; **474**: 21–29.
20. McDonald, T.J., Jornvall, J., Nilsson, G., Vagne, M., Ghatei, M., Bloom, S.R., and Mutt, V. Characterization of a gastrin-releasing peptide from porcine non-antral gastric tissue. *Biochem. Biophys. Res. Comm.*, 1979; **90**: 227–233.
21. Minamino, N., Kangawa, K., and Matsuo, H. Neuromedin B: A novel bombesin-like peptide identified in porcine spinal cord. *Biochem. Biophys. Res. Comm.*, 1983; **114**: 541–548.
22. Benya, R.V., Wada, E., Battey, J.F., Fathi, Z., Wang, L.H., Mantey, S.A., Coy, D.H., and Jensen, R.T. Neuromedin B receptors retain functional expression when transfected into BALB 3T3 fibroblasts: Analysis of binding, kinetics, stoichiometry, modulation by guanine nucleotide-binding proteins, and signal transduction and comparison with natively expressed receptors. *Mol. Pharmacol.*, 1992; **42**: 1058–1068.
23. Benya, R.V., Fathi, Z., Pradhan, T., Battey, J.F., Kusui, T., and Jensen, R.T. Gastrin-releasing peptide receptor-induced internalization, down-regulation, desensitization and growth: Possible role of cAMP. *Mol. Pharmacol.*, 1994; **46**: 235–245.
24. Battey, J.F., Way, J.M., Corjay, M.H., Shapira, H., Kusano, K., Harkins, R., Wu, J.M., Slattery, T., Mann, E., and Feldman, R.I. Molecular cloning of the bombesin/gastrin-releasing peptide receptor from Swiss 3T3 cells. *Proc. Natl. Acad. Sci. USA*, 1991; **88**: 395–399.
25. Wada, E., Way, J., Shapira, H., Kusamo, K., Lebacq-Verheyden, A.M., Coy, D., Jensen, R.T., and Battey, J. cDNA cloning, characterization and brain region-specific expression of a neuromedin-B preferring bombesin receptor. *Neuron,* 1991; **6**: 421–430.
26. Fathi, Z., Corjay, M.H., Shapira, H., Wada, E., Benya, R., Jensen, R., Viallet, J., Sausville, E.A., and Battey, J.F. BRS-3: Novel bombesin receptor subtype selectively expressed in testis and lung carcinoma cells. *J. Biol. Chem.*, 1993; **268**: 5979–5984.
27. Ryan, R.R., Weber, H.C., Mantey, S.A., Hou, W., Hilburger, M.E., Pradhan, T.K., Coy, D.H., and Jensen, R.T. Pharmacology and intracellular signaling mechanisms of the native human orphan receptor BRS-3 in lung cancer cells. *J. Pharmacol. Exp. Ther.*, 1998a; **287**: 366–380.
28. Ryan, R.R., Weber, H.C., Hou, W., Sainz, E., Mantey, S.A., Battey, J.F., Coy, D.H., and Jensen, R.T. Ability of various bombesin receptor agonists and antagonists to alter intracellular signaling of the human orphan receptor BRS-3. *J. Biol. Chem.*, 1998b; **273**: 13613–13624.
29. Mantey, S.A., Weber, H.C., Sainz, E., Akeson, M., Ryan, R.R., Pradhan, T.K., Searles, R.P., Spindel, E.R., Battey, J.F., Coy, D.H., and Jensen, R.T. Discovery of a high affinity radioligand for the human orphan receptor, bombesin

receptor subtype 3, which demonstrates that it has a unique pharmacology compared with other mammalian bombesin receptors. *J. Biol. Chem.,* 1997; **272:** 26062–26071.

30. Pradhan, T.K., Katsuno, T., Taylor, J.E., Kim, S.H., Ryan, R.R., Mantey, S.A., Donohue, P.J., Weber, H.C., Sainz, E., Battey, J.F., Coy, D.H., and Jensen, R.T. Identification of a unique ligand which has high affinity for all four bombesin receptor subtypes. *Eur. J. Pharmacol.,* 1998; **343:** 275–287.

31. Reubi, J.C., Wenger, S., Schumuckli-Maurer, J., Schaer, J.C., and Gugger, M. Bombesin receptor subtypes in human cancers: Detection with the universal radoligand (125)I-[D-TYR(6), beta-ALA(11),PHE(13), NLE(14)] bombesin(6–14). *Clin. Cancer Res.,* 2002; **8:** 1139–1146.

32. Reubi, J.C. Peptide receptors as molecular targets for cancer diagnosis and therapy. *Endocr. Rev.,* 2003; **24:** 389–427.

33. Moody, T.W., Chan, D., Fahrenkrug, J., and Jensen, R.T. Neuropeptides as autocrine growth factors in cancer cells. *Curr. Pharm. Des.,* 2003a; **9:** 495–509.

34. Schally, A.V., and Nagy, A. New approaches to treatment of various cancers based on cytotoxic analogs of LHRH, somatostatin and bombesin. *Life Sci.,* 2003; **72:** 2305–2320.

35. Szereday, Z., Schally, A.V., Nagy, A., Plonowski, A., Bajo, A.M., Halmos, G., Szepeshazi, K., and Groot, K. Effective treatment of experimental U-87MG human glioblastoma in nude mice with a targeted cytotoxic bombesin analogue, AN-215. *Br. J. Cancer,* 2002; **86:** 1322–1327.

36. Moody, T.W., Mantey, S.A., Pradhan, T., Schumann, R., Nakagawa, A., Martinez, A., Fusilier, J., Coy, D.H., and Jensen, R.T. Development of high affinity camptothecin-bombesin conjugates which have targeted cytotoxicity for bombesin receptor-containing cells. *J. Biol. Chem.,* 2004; in publication.

37. Moody, T.W., Zia, F., Venugopal, R., Fagarasan, M., Oie, H., and Hu, V. GRP receptors are present in non small cell lung cancer cells. *J. Cell. Biochem. Suppl.,* 1996; **24:** 247–256.

38. Fuselier, J.A., Sun, L., Woltering, S.N., Murphy, W.A., Vasilevich, N., and Coy, D.H. An adjustable release rate linking strategy for cytotoxic peptide conjugates. *Bioorg. Med. Chem. Lett.,* 2003; **13:** 799–803.

39. Cheng, Y., and Prusoff, W.H. Relationship between the inhibition constant (Ki) and the concentration of inhibitor which causes 50 percent inhibition (I_{50}) of an enzymatic reaction. *Biochem. Pharmacol.,* 1973; **22:** 3099–3108.

40. Marchese, A., Chen, C., Kim, Y.M., and Benovic, J.L. The ins and outs of G protein-coupled receptor trafficking. *Trends Biochem. Sci.,* 2003; **28:** 369–376.

41. Grady, E.F., Slice, L.W., Brant, W.O., Walsh, J.H., Payan, D.G., and Bunnett, N.W. Direct observation of endocytosis of gastrin releasing peptide and its receptor. *J. Biol. Chem.,* 1995; **270:** 4603–4611.

42. Slice, L.W., Yee, H.F., Jr., and Walsh, J.H. Visualization of internalization and recycling of the gastrin releasing peptide receptor-green fluorescent protein chimera expressed in epithelial cells. *Receptors Channels,* 1998; **6:** 201–212.

43. Heuser, J.E., and Anderson, R.G. Hypertonic media inhibit receptor-mediated endocytosis by blocking clathrin-coated pit formation. *J. Cell Biol.,* 1989; **108:** 389–400.

44. Wang, L.H., Rothberg, K.G., and Anderson, R.G. Mis-assembly of clathrin lattices on endosomes reveals a regulatory switch for coated pit formation. *J. Cell. Biol.,* 1993; **123:** 1107–1117.

45. Schiller, J.H., Kim, K., Hutson, P., DeVore, R., Glick, J., Stewart, J., and Johnson, D. Phase II study of topotecan in patients with extensive stage small-cell carcinoma of the lung: An Eastern Cooperative Oncology Group Trial. *J. Clin. Oncol.,* 1996; **14:** 2345–2352.

10

Developmental Milestones in the Newborn Mouse

Joanna M. Hill, Maria A. Lim,
and Madeleine M. Stone

Abstract

The need for a simple method of examining the early postnatal development of mouse models of human neurodevelopmental disorders has become evident. The following method for evaluating the developmental milestones of newborn mice allows for fast throughput of large numbers of mice in a battery of tests that examines weight gain and the reflexes and coordinated movements that are expressed at differing periods throughout the first 21 days of life. Sophisticated equipment is not required, and the measures focus on the day of first appearance of a developmental sign, reflex, or coordinated movement.

Key Words: development; mouse; ontogeny; milestones; neurobehavioral; postnatal; neonatal; neurodevelopment; reflex; locomotion.

Introduction

Mice are among the most widely used organisms to model human disorders and, like human infants, are altricial and exhibit a postnatal appearance of significant ontogenetic events of the nervous system. This provides an opportunity to model human neurodevelopmental disorders that have been characterized by growth restriction and delays in the appearance of developmental milestones. In the mouse nervous system, the early postnatal period includes the important elements of brain growth spurt, synaptogenesis, myelination, gliogenesis, apoptosis, and the appearance of most sensory and motor abilities (see review, *(1)*). Therefore, the undeveloped nature of newborn mice provides the opportunity to track the ontogeny of the nervous system through examination of the development of reflexes, muscular

From: *Neuromethods, Vol. 39: Neuropeptide Techniques*
Edited by: I. Gozes © Humana Press Inc., Totowa, NJ

strength, coordination, and the appearance of postnatal developmental milestones.

The technique described here is a modification of our earlier methods developed for assessing the postnatal neurobehavioral development of newborn mice *(2)* and rats *(3,4)*. The testing procedures are derived from those used by Fox *(5)* and Altman and Sudarshan *(6)*. The method incorporates a range of reflex tests, as indicators of neurologic development *(5)*, as well as tests of strength and motor development *(6)*. The procedure is designed to allow fast throughput so that several litters can be examined daily within a relatively short period of time. In addition, the tests are simple, are easy to measure, and require little equipment. The outcomes of the tests focus on the time to accurately perform, or respond to, a stimulus or posture and the first day of successful performance. The use of these parametric measures allows statistical analysis with analysis of variance. Due to litter effects, the means of the litters are used in statistical comparisons, not the data from individual pups (see Ref. 7). For this reason, the ideal experimental design includes a large number of litters with a sampling of mouse pups from each litter.

Behavioral testing is best performed with observers working in pairs. The presence of the primary observer with a second experienced observer provides consistency of observations over time and allows one person to manipulate the mouse and the other to record the time of performance with the stopwatch.

Procedure

On the day of birth, the number of pups in the litter is counted and each pup is weighed. Since growth rate can be influenced by the size of the litter, large litters are reduced to a maximum of 8 or 9 pups and, where possible, excess pups are added to small litters to bring their numbers up to a desirable size. Beginning at postnatal day 1 (P1), mouse pups are examined daily for the acquisition of developmental milestones. Testing is performed at the same time each day. The battery of tests provides an assessment of development throughout the neonatal period because the behaviors measured are expressed at differing periods throughout the first 21 days of life *(2)*. Each day, the pup is evaluated on specific tests that are appropriate for the age of the pup and the data are recorded on the data sheet (Fig. 1). When possible, testing begins 3 days

Developmental Milestones

Mother ID #_____ Pup Tattoo #_____ D.O.B._____ Sex:_____Treatment/Genotype_____

Behavior	Postnatal Days																					
	0	1	2	3	4	5	6	7	8	9	10	11	12	13	14	15	16	17	18	19	20	21
Weight																						
Surface Righting																						
Negative Geotaxis																						
Cliff Aversion																						
Rooting																						
Forelimb Grasp																						
Auditory Startle																						
Ear Twitch																						
Open Field																						
Eye Opening																						
Air Righting																						

Test daily until the response has been positive for two consecutive days

= Days of Observation

Comments:

Fig. 1. Data sheet for developmental milestones. Neonatal mouse pups are examined daily during postnatal days 1 to 21 for performance on a battery of developmental tests assessing strength, coordination, and the appearance of reflexes and developmental milestones. The shaded areas identify the tests to be performed on each day. For most measures, the day of first performance is recorded, and testing continues daily until the response has been observed for two consecutive days.

before the behavior was previously seen to appear in control NIH Swiss and C57BL/6 mouse pups.

Prior to daily testing, the cage housing the mother and her pups is moved into the testing room, and they are allowed to habituate to the room for 1 h before testing begins. After habituation, the mother is temporarily moved into a clean cage while the pups are removed from the home cage and placed with a small amount of the home cage nesting material in a small bowl or weigh boat that is positioned on a heating pad set at 37 °C (Fig. 2). For the first 5 postnatal days, a lamp with a single 60-watt bulb is also placed over the mice to provide heat from above, as well as from the heating pad below.

On P1, each mouse pup is weighed (Fig. 3) and tested for surface righting, negative geotaxis, cliff aversion, and rooting (see below). Following testing on P1, each mouse is tattooed (Fig. 4A) (Ketchum

Fig. 2. To keep the mouse pups warm during the testing period, the pups are placed with a small amount of the home cage nesting material in a small bowl or weigh boat that is placed on a heating pad set at 37 °C. Also, for the first 5 postnatal days, a lamp with a single 60-watt bulb provides heat from above. Each mouse is returned to its mother following testing.

Fig. 3. On the day of birth, and prior to the testing of the mouse each day, the pups are weighed in a small weigh boat on a top-loading balance.

Fig. 4. To identify individual pups, they are tattooed on P1 according to the AIMS Pup Tattoo Identification System (Budd Lake, NJ). A. A thin needle is dipped into a small amount of green ink (Ketchum Animal Tattoo Ink, Lake Luzerne, NY). B. The needle is introduced under the skin of the paw.

Tattoo Ink, Lake Luzerne, NY) with a small dot in the center of the paw (Fig. 4B) following the AIMS Pup Tattoo Identification System (Budd Lake, NJ) and examined to determine its gender. The data for each mouse are recorded on a separate sheet shown in Fig. 1. The shaded areas on the data sheet indicate the behaviors to be tested on each day. The data sheet also contains areas for recording identifying information about the pup, including the mother's ID number, the pup tattoo number, date of birth, gender, and genotype or treatment group.

The tests are performed on the pups following the schedule outlined on the data sheet. Unless otherwise specified, tests are performed on the smooth side of a sheet of laboratory bench paper that can be cleaned or replaced between litters. Tests are repeated daily until the pup meets criterion for two consecutive days.

Surface Righting

(Labyrinthine and body righting mechanisms, strength, and coordination) The mouse pup is held gently on its back, with the investigator's two fingers holding either side of the head and two fingers holding the hind quarters (Fig. 5A) on a smooth sheet of plastic. The mouse pup is released and the time in seconds for the pup to flip over onto its abdomen with all four paws touching the surface of the table measured (Fig. 5B). If the mouse does not respond within 30 s, the test is terminated. Surface righting is measured once daily until the mouse can right itself in less than 1 s for two consecutive days.

Negative Geotaxis

(Labyrinthine reflex and body righting mechanisms, strength, and coordination) The mouse is placed head down on a square of screen set at an angle of 45° (Figs. 6A and B). The time in seconds for the pup to turn 180° to the "head up" position (Figs. 6C and D) is recorded. If the pup loses footing and slips on the screen, the test can be repeated once. If the mouse does not respond within 30 s, the test is terminated. The test is repeated daily until it is performed correctly in fewer than 30 s for two consecutive days.

Cliff Aversion

(Labyrinthine reflex, strength, and coordination) The mouse pup is positioned on the edge of a small box with a smooth surface with the digits of the forepaws and the snout hanging over the edge of the box (Fig. 7A). The time in seconds for the pup to turn (Fig. 7B) and begin to crawl away from the edge (Fig. 7C) is measured. If the pup loses footing and slips off the box, the test can be repeated once. If the mouse does not respond within 30 s, the test is terminated. The test is repeated daily until it is performed correctly in fewer than 30 s for two consecutive days.

Rooting

(Tactile reflex and motor coordination) The cotton tip of an applicator (Salon Manufacturing Company, Solon, ME) is pulled out and the end is twisted to form a fine filament. The filament is

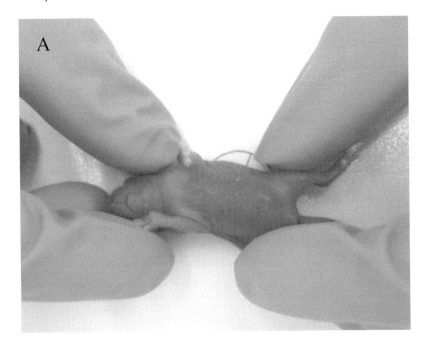

Fig. 5. Surface righting. A. The pup is gently held on its back and then released. B. The response is timed until the mouse pup flips over onto its abdomen with all four paws touching the table. If the mouse has not righted in 30 s, the test is terminated.

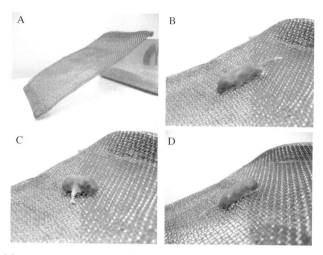

Fig. 6. Negative geotaxis. A. Negative geotaxis screen set at a 45° angle. B. The mouse is placed on the screen head down and released. C. The mouse begins to turn around. D. The response is timed until the mouse has turned enough that its nose is facing up. If the mouse has not turned in 30 s, the test is terminated.

slowly and gently stroked three times from front to back along the side of the pup's head (Fig. 8A). Rooting occurs if the pup moves its head toward the filament (Fig. 8B). If there is no response to brushing the filament on one side of the head, the test is repeated once on the other side of the head. Testing continues daily until the pup responds correctly for two consecutive days.

Forelimb Grasp

(Strength) A pup is held with its forepaws resting on a small rod suspended over shavings 5 cm deep until it grasps the rod (Fig. 9A). The pup is then released, and the length of time the mouse remains gripping the rod is measured (Fig. 9B). If the pup falls off immediately, the test can be repeated once. The test is repeated daily until the mouse can remain gripping the rod for a minimum of 1 s for two consecutive days.

Auditory Startle

(Auditory reflex) The mouse pup is placed on the laboratory bench, and its reaction to a handclap at a distance of 10 cm is

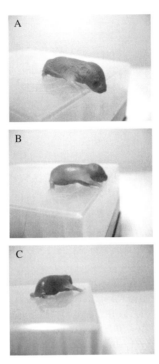

Fig. 7. Cliff aversion. A. The mouse is placed with the digits of its forepaws and nose hanging over the edge of the cliff. B. The mouse begins to back away and turn from the cliff. C. The response is timed until both paws and its nose are no longer protruding over the cliff. If the mouse does not have both paws and its nose removed from the cliff in 30 s, the test is terminated.

recorded (Fig. 10). The mouse responds with a quick involuntary jump. The test is repeated daily until the pup responds correctly for two consecutive days.

Ear Twitch

(Tactile reflex) The cotton tip of an applicator (Salon Manufacturing Company) is pulled out and the end is twisted to form a fine filament. The filament is gently brushed against the tip of the ear three times (Fig. 11). The mouse responds by flattening the ear against the side of the head. The test is repeated daily until the mouse responds correctly for two consecutive days.

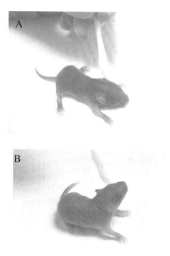

Fig. 8. Rooting. A. A fine filament of cotton is gently stroked three times along the side of the head of the pup. B. The pup responds by turning the head toward the filament.

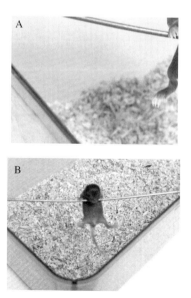

Fig. 9. Forelimb grasp. A. The pup is held against a wire suspended over a cage with 5 cm of shavings. B. After release, the length of time the pup can remain suspended is recorded.

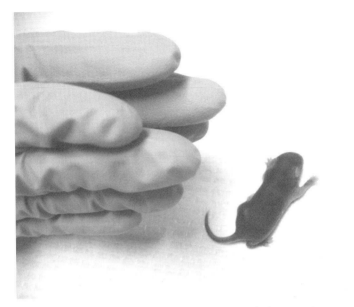

Fig. 10. Auditory startle. The presence of the startle response of the mouse to a clapping sound is noted.

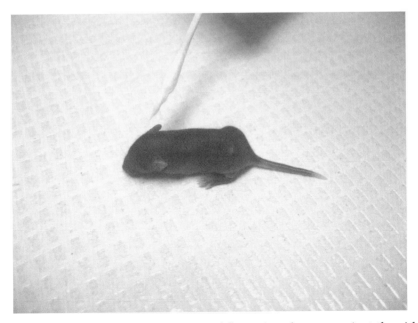

Fig. 11. Ear twitch. The response of flattening the ear against the side of the head is noted following brushing the tip of the ear with a fine cotton filament.

Open Field Traversal

(Locomotion and the extinguishing of pivoting behavior) The mouse is placed on a sheet of plastic in the center of a circle 13 cm in diameter (Fig. 12A). The length of time the mouse takes to move out of the circle (Fig. 12B) is recorded. If the mouse does not respond in 30 s, the test is terminated. The mouse is tested daily until it responds in fewer than 30 s for two consecutive days.

Eye Opening

(Developmental milestone) The mouse pup is examined daily to determine the first day that both eyes are open.

Air Righting

(Labyrinthine and body righting mechanisms and coordination) The mouse pup is held upside down by the investigator, with two fingers holding either side of the head and two fingers holding the hind quarters (Fig. 13A) approximately 10.5 cm over a cage containing 5 cm of shavings. The pup is released, and its position upon landing on the shavings is examined. The first day that the pup released upside-down turns right side up and lands on all four paws on a bed of shavings (Fig. 13B) is recorded. The test is repeated daily until the pup lands on all four paws for two consecutive days.

Statistical Analysis and Alternate Methods of Data Collection

The data are analyzed using analysis of variance, with repeated measures where appropriate, followed by contrasts between controls and all other treatment groups. Because of potential litter effects, the data used in analysis are the means of the litters for that variable *(7)*.

The data collection method allows for greater depth in analyzing behaviors than described above. For example, the method described here defines arbitrary criteria for performance: 1 s for surface righting, 1 s for forelimb grasp, and 30 s for open field traversal, and examines the first day the behavior appears. However, in previous studies, the speed performance of many of these behaviors was examined over time *(2–4)*. Alternately,

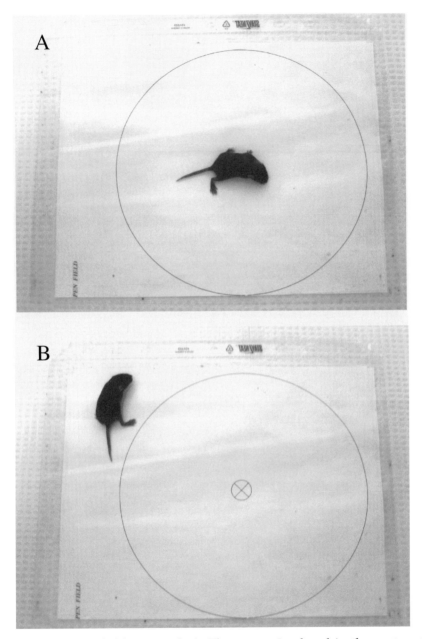

Fig. 12. Open field traversal. A. The mouse is placed in the center of a 13-cm circle. B. The time taken to move off the circle is recorded. If the mouse does not move off the circle in 30 s, the test is terminated.

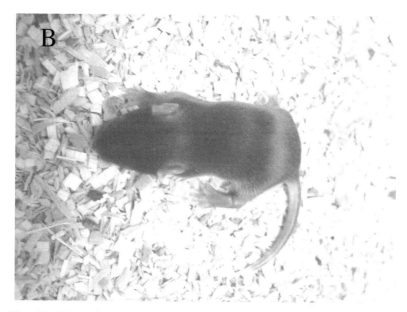

Fig. 13. Air righting. A. The mouse is held upside down over a cage containing 5 cm of shavings. B. After the mouse is released, the response of landing right side up with all four feet on the shavings is noted.

the time taken to perform a behavior on a particular day was analyzed *(2–4)*. In addition, Fox *(5)* describes developmental data with a numerical scoring method including both the strength of the response, from 0 to 9, and the reliability of the response—the percentage of the total number of offspring in which the response is seen as a maximal effect. The chronological development of reflexes is also graphed. These alternate methods may prove beneficial for pulling apart small differences in performance among genotypes or treatment groups.

Discussion

Figure 14 and Table 1 illustrate data from Hill *(8)*, collected and analyzed as described above. Vasoactive intestinal peptide (VIP) has an important role in the regulation of growth and development during early postimplantation embryogenesis (see review, *(8)*). VIP and the related neuropeptide, pituitary adenylate cyclase activating peptide (PACAP), act on the same receptors and are known to have similar actions in some systems. The study examined the offspring of pregnant mice that had been

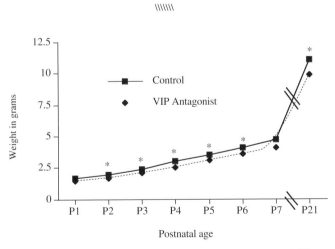

Fig. 14. Weight in grams of postnatal mouse pups. VIP-antagonist-treated pups experienced blockage of VIP with a VIP hybrid antagonist *(13)* (30 µg/day), injected during embryogenesis, as previously described *(2)*. P = postnatal day. VIP-antagonist-treated pups were significantly growth restricted from P2 to P21, *$p < 0.05$. (Reprinted with permission *(8)*.)

Table 1. Neonatal Developmental Milestones. Neonatal mice were tested daily for the first 21 days of life for a number of reflexes and developmental signs.

Behavior	Control	VIP antagonist[#]	PACAP antagonist[##]	F(p)
Weight in grams P2	1.9±0.03	1.7±0.03***	1.9±0.23	21 (0.0001)
Surface Right first day <1 sec	5.3±0.33	8.4±0.26***	5.4±0.24	23.8 (0.0001)
Negative Geotaxis first day	3.8±0.19	3.8±0.35	3.0±0.21	3.47 (0.051)
Cliff Aversion first day	4.3±0.25	4.2±0.34	3.8±0.13	1.13 (0.34)
Rooting first day	4.6±0.2	4.6±0.1	4.5±0.3	0.11 (0.90)
Grasping first day > 1 sec	8.3±0.3	10.3±0.1***	7.4±0.4	15.8 (0.0001)
Open field first day	14.7±0.4	18.5±0.3***	14.2±0.3	34.1 (0.0001)
Auditory Startle first day	10.8±0.3	11.7±0.1*	10.5±0.3	3.7 (0.042)
Ear Twitch first day	10.0±0.1	10.5±0.2	9.7±0.2	2.7 (0.09)
Eye Opening first day	13.4±0.1	13.6±0.2	13.7±0.1	2.8 (0.08)
Air righting first day	10.9±0.1	11.3±0.3	11.1±0.2	1.08 (0.35)

[#]VIP antagonist = VIP hybrid antagonist (10) injected embryonic days 8–11, 30 µg/day.
[#]PACAP antagonist = PACAP 6–38, injected embryonic days 8–11, 30 $UP\mu$g/day.
*= p<0.05; *** = p<0.0001.
Reprinted with permission (8).

treated with antagonists to VIP or PACAP during early embryogenesis. All mice were born within a two-day period, and all experimental groups were studied concurrently. The offspring of females that had been treated with the VIP antagonist exhibited delays in five measures compared with the offspring of untreated mothers. PACAP-antagonist-treated pups did not differ from control pups in any measures. The results demonstrated that VIP and PACAP had different roles at this developmental stage and further indicated that the effect of blocking VIP action in this manner was specific.

The offspring of VIP-antagonist-treated mice weighed less at birth and continued to weigh significantly less than control animals through the entire study period of 21 days (Fig. 14). Lower birth weight and reduced stature are associated with human neurological disorders, including Down syndrome *(9)*, and can indicate a prenatal developmental defect that cannot be reversed nutritionally.

Surface righting was delayed by 3 days in the VIP-antagonist-treated pups, grasping by 2 days, open field traversal by 4 days, and auditory startle by 1 day, indicating that, by several measures, the postnatal development of reflexes and strength was retarded in these mice. Delays in reflexes can indicate that the underlying neuronal circuits are developing at a slower rate than in controls. However, reflexes such as righting require the concurrent development of strength and coordination, and a delay in their appearance may not indicate retarded development of the nervous system so much as muscle weakness or delayed muscular development. However, as there was a delay in the appearance of auditory startle, which does not require coordinated movement, the data suggest that neuronal development is retarded. The fact that the milestone of eye opening and the reflexes of rooting and ear twitch did not differ between the VIP-antagonist-treated pups and control pups suggests that the method was also able to differentiate between the developmental rates of different systems. Locomotion develops gradually during the first 2 weeks of life in this altricial species, and the forelimbs are involved before the hind limbs in development, resulting in pivoting behavior *(6)*. Open field traversal examines the extinguishing of pivoting behavior and the development of straight-line walking, reflecting the rostrocaudal gradient of maturation of limb coordination *(6)*.

Interestingly, the significant delays observed in the acquisition of developmental milestones in mice deficient in apolipoprotein E, a molecule associated with the etiology of Alzheimer's disease, and rats undergoing hypoxia and cholinergic blockage as models of developmental retardation have been reversed with treatment with a VIP agonist *(10,11)*. These studies underscore the value of following the progress in achievement of developmental milestones in research animals.

Conclusions

The method outlined here provides a useful technique for rapidly assessing the early postnatal development of newborn mice. There is a growing need for simple methods to measure these parameters, as the increasing importance in assessing the developmental profile of genetically modified mice has recently been recognized *(12)*. Analysis of the progress of neonatal transgenic and knockout mice in acquiring developmental milestones will reveal developmental abnormalities and will permit a better understanding of the basis for adult anomalies through serving as animal models of neurodevelopmental diseases.

Acknowledgments

This research was supported by the Intramural Research Program of the NIH, NIMH, and the US-Israel Binational Science Foundation. The authors gratefully acknowledge the expert technical assistance of Daniel Abebe in the development of the method. This is an invited review for a book on neuropeptide techniques edited by Professor Illana Gozes.

References

1. Branchi, I., Bichler, Z., Berger-Sweeney, J., and Ricceri, L. Animal models of mental retardation: From gene to cognitive function. *Neurosci. Biobehav. Rev.,* 2003; **27**: 141–153.
2. Wu, J.Y., Henins, K.A., Gressens, P., Gozes, I., Fridkin, M., Brenneman, D.E., and Hill, J.M. Neurobehavioral development of neonatal mice following blockade of VIP during the early embryonic period. *Peptides,* 1997; **18**: 1131–1137.
3. Hill, J.M., Gozes, I., Hill, J.L., Fridkin, M., and Brenneman, D.E. Vasoactive intestinal peptide antagonist retards the development of neonatal behaviors in the rat. *Peptides,* 1991; **12**: 187–192.
4. Hill, J.M., Mervis, R.F., Avidor, R., Moody, T.W., and Brenneman, D.E. HIV envelope protein-induced neuronal damage and retardation of behavioral development in rat neonates. *Brain Res.,* 1993; **603**: 222–233.
5. Fox, W.M. Reflex-ontogeny and behavioral development of the mouse. *Anim. Behav.,* 1965; **13**: 234–241.
6. Altman, J., and Sudarshan, K. Postnatal development of locomotion in the laboratory rat. *Anim. Behav.,* 1975; **23**: 896–920.

7. Zorrilla, E.P. Multiparous species present problems (and possibilities) to developmentalists. *Dev. Psychobiol.*, 1997; **30**: 141–150.
8. Hill, J.M. Vasoactive intestinal peptide in neurodevelopmental disorders: Therapeutic potential. *Curr. Pharm. Design* (in publication).
9. Epstein, C.J. Down syndrome (trisomy 21). In *The Metabolic Basis of Inherited Disease*, C.R. Shriver, A.L. Beaudet, W.S. Sly, and V. Valle, eds. McGraw-Hill, New York, 1989, pp. 291–326.
10. Gozes, I., Bachar, M., Bardea, A., Davidson, A., Rubinraut, S., Fridkin, M., and Giladi, E. Protection against developmental retardation in apolipoptrotein E-deficient mice by a fatty neuropeptide: Implications for early treatment of Alzheimer's disease. *J. Neurobiol.*, 1997; **33**: 329–342.
11. Gozes, I., Bachar, M., Bardea, A., Davidson, A., Rubinraut, S., and Fridkin, M. Protection against developmental deficiencies by a lipophilic VIP analogue. *Neurochem. Res.*, 1998; **23**: 689–693.
12. Branchi, I., and Ricceri, L. Transgenic and knock-out mouse pups: The growing need for behavioral analysis. *Genes Brain Behav.*, 2002; **1**: 135–141.
13. Gozes, I., McCune, S.K., Jacobson, L., Warren, D., Moody, T.W., Fridkin, M., and Brenneman, D.E. An antagonist to vasoactive intestinal peptide affects cellular functions in the central nervous system. *J. Pharmacol. Exp. Ther.*, 1991; **257**: 959–966.

Index

151

Printed in the United States of America